MESSAGING FOR ENGINEERING

FROM RESEARCH TO ACTION

Committee on Implementing Engineering Messages

NATIONAL ACADEMY OF ENGINEERING
OF THE NATIONAL ACADEMIES

THE NATIONAL ACADEMIES PRESS
Washington, D.C.
www.nap.edu

NATIONAL ACADEMIES PRESS 500 Fifth Street, NW Washington, DC 20001

NOTICE: This publication has been reviewed according to procedures approved by the National Academy of Engineering report review process. Publication of signed work signifies that it is judged a competent and useful contribution worthy of public consideration, but it does not imply endorsement of conclusions or recommendations by the National Academy of Engineering. The interpretations and conclusions in such publications are those of the authors and do not purport to present the views of the council, officers, or staff of the National Academy of Engineering.

This material is based on work supported by the National Science Foundation under Grant No. EEC-0957578. Any opinions, findings, and conclusions or recommendations expressed in this material are those of the author(s) and do not necessarily reflect the views of the National Science Foundation.

Library of Congress Cataloging-in-Publication Data

Messaging for engineering : from research to action / Committee on Implementing Engineering Messages.
pages cm
"This publication builds on the 2008 NAE publication, Changing the conversation : messages for improving public understanding of engineering"—Preface.
Includes bibliographical references.
ISBN 978-0-309-26180-7 (paperback) — ISBN 978-0-309-26181-4 (pdf) 1. Communication in engineering. 2. Engineering—Social aspects—United States. 3. Engineers—Public relations—United States. I. National Academy of Engineering. Committee on Implementing Engineering Messages. II. National Academy of Engineering. Committee on Public Understanding of Engineering Messages. Changing the conversation.
TA158.5.M47 2013
620.001'4—dc23

2013018880

Copies of this report are available from the National Academies Press, 500 Fifth Street, NW, Keck 360, Washington, DC 20001; (888) 624-8373 or (202) 334-3313; *www.nap.edu.*

For more information about the National Academy of Engineering, visit the NAE home page at *www.nae.edu.*

Printed in the United States of America

THE NATIONAL ACADEMIES

Advisers to the Nation on Science, Engineering, and Medicine

The **National Academy of Sciences** is a private, nonprofit, self-perpetuating society of distinguished scholars engaged in scientific and engineering research, dedicated to the furtherance of science and technology and to their use for the general welfare. Upon the authority of the charter granted to it by the Congress in 1863, the Academy has a mandate that requires it to advise the federal government on scientific and technical matters. Dr. Ralph J. Cicerone is president of the National Academy of Sciences.

The **National Academy of Engineering** was established in 1964, under the charter of the National Academy of Sciences, as a parallel organization of outstanding engineers. It is autonomous in its administration and in the selection of its members, sharing with the National Academy of Sciences the responsibility for advising the federal government. The National Academy of Engineering also sponsors engineering programs aimed at meeting national needs, encourages education and research, and recognizes the superior achievements of engineers. Dr. Charles M. Vest is president of the National Academy of Engineering.

The **Institute of Medicine** was established in 1970 by the National Academy of Sciences to secure the services of eminent members of appropriate professions in the examination of policy matters pertaining to the health of the public. The Institute acts under the responsibility given to the National Academy of Sciences by its congressional charter to be an adviser to the federal government and, upon its own initiative, to identify issues of medical care, research, and education. Dr. Harvey V. Fineberg is president of the Institute of Medicine.

The **National Research Council** was organized by the National Academy of Sciences in 1916 to associate the broad community of science and technology with the Academy's purposes of furthering knowledge and advising the federal government. Functioning in accordance with general policies determined by the Academy, the Council has become the principal operating agency of both the National Academy of Sciences and the National Academy of Engineering in providing services to the government, the public, and the scientific and engineering communities. The Council is administered jointly by both Academies and the Institute of Medicine. Dr. Ralph J. Cicerone and Dr. Charles M. Vest are chair and vice chair, respectively, of the National Research Council.

www.national-academies.org

COMMITTEE ON IMPLEMENTING ENGINEERING MESSAGES

ELLEN KULLMAN (*Cochair*), E. I. du Pont de Nemours and Company, Wilmington, DE
CHARLES M. VEST (*Cochair*), National Academy of Engineering, Washington, DC
G. WAYNE CLOUGH, Smithsonian Institution, Washington, DC
LESLIE COLLINS, National Engineers Week Foundation, Alexandria, VA
DON P. GIDDENS, Georgia Institute of Technology, Atlanta
ROBERT HOFFMAN, Hoffman and Partners LLC, Boston, MA
RAY O. JOHNSON, Lockheed Martin Corporation, Bethesda, MD
VIRGINIA KRAMER, Keiler & Company, Farmington, CT
IRVING PRESSLEY MCPHAIL, National Action Council for Minorities in Engineering, White Plains, NY
E. JAMES PRENDERGAST, Institute of Electrical and Electronics Engineers, Piscataway, NJ
BETTY SHANAHAN, Society of Women Engineers, Chicago, IL

Staff

GREG PEARSON, Senior Program Officer, National Academy of Engineering
MARIBETH KEITZ, Senior Program Associate, National Academy of Engineering

PREFACE

This report is the final product of the Committee on Implementing Engineering Messages of the National Academy of Engineering (NAE). The project overseen by the committee had three objectives: (1) develop an online "toolkit" containing messaging-related resources, community-building applications, and other resources to support the goal of promoting broader use of new messages for improving the public understanding of engineering; (2) facilitate dialogue between organizations that have developed implementation strategies for the new engineering messages and influential stakeholders in the engineering community that have not yet implemented the messages; and (3) create an "action plan" to guide adoption and use of the online toolkit and encourage coordinated outreach to the public by the broader engineering community. The committee's report fulfills the project's third objective.

This publication builds on the 2008 NAE publication *Changing the Conversation: Messages for Improving Public Understanding of Engineering. Changing the Conversation* (CTC) presented the results of a research-based effort to develop and test new, more effective ways of communicating to the public about engineering. The new messages recast engineering as inherently creative and concerned with human welfare, as well as an emotionally satisfying calling. The CTC report stimulated considerable interest among segments of the engineering community, and some organizations adopted the project's messages in their outreach. Overall, however, the report's impact has fallen short of its potential to galvanize action by the broader engineering community. Our committee's effort is intended to help remedy that situation.

This report provides background related to engineering messaging, reports on progress made in implementing the CTC messages, and suggests specific steps that can be taken by each of the major players in the engineering community to continue and build on the implementation that has taken place. As we note in the report, like any rebranding effort, creating more accurate and positive perceptions of engineering is a long-term proposition. Through the work of the NAE and other organizations, the engineering community has the tools in hand to make considerable progress. The information in this document, we hope, will provide incentive for meaningful action.

Ellen Kullman, *Cochair*
Chair of the Board and CEO
E. I. du Pont de Nemours
and Company

Charles M. Vest, *Cochair*
President
National Academy of
Engineering

ACKNOWLEDGMENTS

This report has been reviewed in draft form by individuals chosen for their diverse perspectives and technical expertise, in accordance with procedures approved by the National Academies. The purpose of the independent review is to provide candid and critical comments to assist the NAE in making its published report as sound as possible and to ensure that the report meets institutional standards for objectivity, evidence, and responsiveness to the study charge. The review comments and draft manuscript remain confidential to protect the integrity of the deliberative process. We wish to thank the following individuals for their review of this report:

James Buczkowski, Ford Motor Company
Norman L. Fortenberry, American Society for Engineering Education
Lisa Guerra, NASA Headquarters
William S. Hammack, University of Illinois, Urbana-Champaign
Thomas G. Loughlin, ASME
Patrick J. Natale, American Society of Civil Engineers
Mary Petryszyn, Raytheon Company

Darryll J. Pines, University of Maryland
Anthony (Bud) Rock, Association of Science-Technology Centers
Jacquelyn F. Sullivan, University of Colorado–Boulder

Although the reviewers listed above provided many constructive comments and suggestions, they were not asked to endorse the views expressed in the report, nor did they see the final draft of the report before its release. The review of this report was overseen by Julia M. Phillips, Sandia National Laboratories. Appointed by the NAE, she was responsible for making certain that an independent examination of this report was carried out in accordance with institutional procedures and that all review comments were carefully considered. Responsibility for the final content of this report rests entirely with the authors and the NAE.

CONTENTS

Appendixes

MESSAGING FOR ENGINEERING

SUMMARY

Messaging for Engineering: From Research to Action is the final report of a project undertaken by the National Academy of Engineering (NAE) with funding from the National Science Foundation. The project goal was to support efforts by the engineering community to communicate more effectively about the profession and those who practice it. It was overseen by a 10-member advisory committee cochaired by Ellen Kullman, chair of the board and CEO of DuPont, and NAE President Charles M. Vest. The committee's report builds on the 2008 NAE publication *Changing the Conversation: Messages for Improving Public Understanding of Engineering* ("the CTC report"), which presented the results of a research-based effort to develop and test new, more effective messages about engineering (Box S-1).

The new messages cast engineering as inherently creative and concerned with human welfare, as well as an emotionally satisfying calling. The 2008 report stimulated considerable interest among segments of the engineering community, and some organizations adopted or adapted the project's messages in their outreach. This report summarizes progress in implementing the CTC messages, but also recognizes that there is potential to galvanize additional action and thus suggests

BOX S-1
Four New Messages for Promoting Engineering

- *Engineers make a world of difference.* From new farming equipment and safer drinking water to electric cars and faster microchips, engineers use their knowledge to improve people's lives in meaningful ways.
- *Engineers are creative problem solvers.* They have a vision for how something should work and are dedicated to making it better, faster, or more efficient.
- *Engineers help shape the future.* They use the latest science, tools, and technology to bring ideas to life.
- *Engineering is essential to our health, happiness, and safety.* From the grandest skyscrapers to microscopic medical devices, it is impossible to imagine life without engineering.

NOTE: Adapted from NAE (2008).

specific steps for major players in the engineering community to continue and build on progress to date. Many of the report's recommendations resulted from discussion at a December 2010 committee workshop that involved several dozen high-level decision makers representing key stakeholder groups in the engineering community.

SAMPLES OF IMPLEMENTATION

Since the 2008 release of the CTC report, a number of institutions have either directly used or adapted its messages and related "taglines." The Society of Women Engineers, for example, reworked all of its print and web-based messaging products to align with the CTC positioning statement and messages. At the University of Colorado–Boulder, the College of Engineering undertook a major rebranding effort that included the redesign of recruiting brochures and posters that now feature CTC messages. Increases in enrollment and retention of minority and women students at the college may in part be attributable to the effect of messaging. The Engineer Your Life website (www.engineeryourlife.org), the centerpiece of a national campaign to encourage college-bound girls to explore engineering, is using some of the CTC messages as well as messages similar in content. The site also links to an Engineer's Pledge on Facebook (www.facebook.com/engineerpledge), which asks engineers to agree to use CTC-infused language when they talk to the public. The Institute of Electrical and Electronics Engineers–USA used the CTC message "Engineers make a world of difference" as the basis of an annual $5,000 scholarship contest for undergraduate students, with the challenge of creating a 90-second video promoting engineering to 11- to 13-year-olds. The National Engineers Week Foundation (NEWF) has used CTC messages and taglines to promote its Future City competition and in ads in *USA Today* announcing its New Faces in Engineering program, which recognizes the work of early-career engineers. And the NAE has spread the CTC messages and taglines by linking them to its Grand Challenges for Engineering project (www.engineeringchallenges.org) and to the nomination process for the Bernard M. Gordon Prize, a $500,000 award that recognizes innovations in engineering and technology education.

In addition, a number of large, well-known companies, including DuPont, Texas Instruments, Cisco, Exxon Mobil, GE, and Lockheed Martin, have produced advertising or recruiting materials that reflect the spirit of *Changing the Conversation.* They focus on the value of engineering to people's lives and on the creativity of the engineering profession, and show that a career in engineering is within reach for many young people who have vision and a desire to solve problems and help people.

CREATION OF ADDITIONAL CTC-RELATED TOOLS

The NAE has also developed tools to help spread the CTC messages, including an interactive website (www.engineeringmessages.org) that provides background on the issue of public understanding of engineering; features nearly 140 examples of engineering messaging efforts; and includes free, downloadable CTC-branded posters, bookmarks, and door hangers. To encourage sharing via social media, nearly every page has tabs for Google+, Facebook, Twitter, and Pinterest. With support from the United Engineering Foundation (www.uefoundation.org; UEF), the NAE also created a Facebook site (www.facebook.com/engineersctc) to complement the CTC website.

LOOKING TO THE FUTURE

As has been clear since before publication of the CTC report, propagating a new image of engineering will require not just websites and marketing campaigns but also the work of many individual proponents who spread the message in one-on-one and group interactions.

To this end, the committee has prepared an action plan that describes (1) basic steps for all segments of the engineering community to help change the conversation about engineering and (2) specific steps for individual segments of the engineering community.

Call to Action: Basic Steps

Organizations and individuals across the engineering community (as well as nonengineers who believe in the importance of engineering) can take several basic actions that will contribute to public understanding of engineering.

- Make explicit use of the words "engineer" and "engineering" and express the CTC positioning more frequently in public communications, such as press releases, radio and television advertising, websites, social media (e.g., Facebook and Twitter), speeches, and personal email signature blocks. This approach can also be incorporated in internal communications to employees or colleagues.
- Engage more fully with the CTC website (www.engineeringmessages.org) and Facebook page (www.facebook.com/engineersctc). These resources provide practical help for effective messaging and opportunities to build a community of practice.

- Use the CTC messages and taglines with a view of the engineering profession as a whole. That said, messaging tuned to individual disciplines may be useful in certain situations, and "cobranding" by combining general and discipline-specific messaging is certainly possible.
- Take into account the importance of reaching girls, African Americans, Hispanics, and American Indians, groups that data show are significantly underrepresented in engineering in the United States.
- Assess the impact of using or adapting the CTC messages. Ideally, such evaluation should be included up front in the design of the messaging effort rather than as an add-on after the fact. As much as possible, measures of impact should include not only "inputs," such as the number of visits to a website, but also "outputs" that reflect changes in attitudes or behavior—for example, students' views of engineering—as a result of exposure to the CTC messages.

Call to Action: Industry

The committee recommends that industry take the following actions to help change the conversation about engineering:

- Increase the number of companies whose corporate identity, recruiting efforts, product advertising, and outreach to the public feature engineers and engineering and use messages and taglines either directly from or comparable to those from *Changing the Conversation.*
- Leverage outreach and messages by collaborating more often with other segments of the engineering community, such as professional societies and engineering colleges. They could also collaborate among themselves through such mechanisms as the Council on Competitiveness, Business Roundtable, and Change the Equation to create consistent, CTC-based messaging for large segments of the US population.

- Support volunteer outreach by employees to interest young people in engineering. At DuPont, for example, more than 100 employees involved in science, technology, engineering, and mathematics (STEM) outreach were trained in the use of the CTC messages; similar training was provided to company staff volunteering at engineering events for the Girl Scouts and for over 20 high schools near the company's headquarters. Industry efforts like these should continue, expand if possible, and be modified to align with the CTC messages.
- Exert a positive influence on diverse segments of society through philanthropic giving. These companies should ask nonprofit organizations that they support and that engage in engineering-related outreach to use the CTC messages.
- Consider investing in public service announcements that project a positive image of engineering consistent with CTC messaging. Such efforts could be tied to television programming that connects to engineering in some way, such as Discovery Channel's Extreme Engineering and MythBusters, the Science Channel's Strip the City, and PBS's Design Squad.

Call to Action: Government Agencies

The committee recommends that government agencies take the following actions to help change the conversation about engineering:

- Incorporate the CTC messages in education and outreach programs, such as NASA's Summer of Innovation, and in all STEM-related government programs that support hands-on experiments and engineering design activities for schools, libraries, scout troops, civic centers, and other organizations.
- Collaborate with other segments of the engineering community to advance the goal of changing the conversation—for example, by working with industry partners in outreach programs or regularly participating in the CTC website.
- Incorporate the CTC materials in training for federally employed engineers who take on speaking and mentoring assignments with students and educators.

- Create incentives for recipients of federal grants and contracts to incorporate CTC messaging in their work.

Call to Action: Engineering Professional Societies

The committee recommends that engineering professional societies take the following actions to help change the conversation about engineering:

- To provide motivation and vision for working together to improve public understanding of engineering, develop and endorse a shared "memorandum of understanding" to guide coordinated use of the CTC messages.
- Educate members about the messages and how to use them. This could be done, for instance, by offering training sessions at society-sponsored conferences and workshops, through webinars (like those sponsored by NEWF, described in Chapter 2), and by including articles and editorials about the CTC project in society publications.
- Conduct outreach to teachers, students, and parents. Society-sponsored teacher conferences and workshops could focus on the theme of *Changing the Conversation*, and materials for K–12 teachers could explain how engineering makes the world better, reflecting one of the main CTC themes. Communications about the engineering profession as a whole should be the main focus, but messages tuned to individual disciplines are also useful.

Call to Action: American Society for Engineering Education

The committee recommends that the American Society for Engineering Education (ASEE) leverage its special connection to engineering educators to broaden their awareness and use of the CTC messages. Specific steps might include the addition of a recurring session at the annual ASEE conference and at the yearly Engineering Deans Council Public Policy Colloquium to review and encourage discussion of efforts to improve engineering messaging in engineering education programs around the country. ASEE could also provide visibility by selecting one program's messaging efforts each year for special recognition and featuring the program in a one-page spread in *Prism* magazine.

Call to Action: National Engineers Week Foundation

The committee recommends that NEWF take the following actions to help change the conversation about engineering:

- Continue to offer web-based, messaging-focused training to engineering student "ambassadors" and other volunteers who do outreach to the community and to K–12 schools.
- Encourage corporate sponsors to incorporate the CTC messages in their public outreach not only during E-Week but throughout the year.

Call to Action: Engineering Schools

The committee recommends that engineering schools take the following actions to help change the conversation about engineering:

- Explain the CTC messaging approach to faculty and staff, describing its rationale and the evidence for its usefulness. This might be done in faculty orientation workshops or other training sessions that lay out explicitly how to use the CTC messages and taglines to shape how students and potential students think about engineering.
- Spread the CTC messages to current and potential students by, for example, incorporating the CTC messages and taglines in the recruiting and outreach programs of the engineering school. Such efforts need not be directed only to high school juniors and seniors and university students who have not yet decided on a major.
- Work with schools of education so that future K–12 teachers are aware of what engineering is and what engineers do, and encourage schools of education to use CTC-based messages.

- Encourage engineering undergraduates to volunteer in K–12 classrooms doing engineering design projects, acting as role models, and using CTC-based messages. This kind of outreach can shape K–12 students' impressions of engineers and engineering.
- Aim to educate students consistent with the image of engineering outlined in *Changing the Conversation.* If K–12 students are attracted to engineering by the opportunity to both engage their creativity and help people, the undergraduate engineering curriculum should reflect those qualities. Some schools may wish to use the Grand Challenges for Engineering (www.engineeringchallenges.org) to inspire students to think about problems whose solutions will make a "world of difference." Engineering students educated in these ways can themselves become ambassadors for spreading the CTC messages.

Call to Action: Science and Technology Centers

The committee recommends that science and technology centers take the following actions to help change the conversation about engineering:

- When designing new exhibits or revising existing ones, incorporate the CTC messages to the extent possible. Exhibits and other programming at science and technology centers can educate the public about engineering in very engaging ways, and this capacity should be leveraged to deliver the CTC messages compellingly and memorably.
- Involve engineers from academia, professional societies, and industry in programs and outreach activities that have an engineering or technology focus.

Call to Action: National Academy of Engineering

The committee recommends that the NAE take the following actions to help change the conversation about engineering:

- Continue to spread ideas from the CTC project and to promote communication and actions that support public understanding of engineering. This report is the most recent such effort.
- Maintain the CTC website and Facebook page until they are deemed no longer useful. This could be done by the NAE alone or in partnership with other engineering organizations.

1 INTRODUCTION

For those in the broad engineering community—those who employ, work with, and/or educate engineers, and engineers themselves—there is no need to explain the importance and value of engineering. They understand that engineers help make the world a better place for all, that they regularly grapple with important societal and environmental issues, and that the engineering process is every bit as creative as composing a symphony or crafting a piece of art.

But the situation outside the engineering community is quite different. Studies have shown that most K–12 students and teachers have a limited appreciation of all the ways that engineering makes their lives better and, furthermore, that they have little understanding of what engineers do or of the opportunities that an engineering education offers (Cunningham et al. 2006; Lachapelle et al. 2012). While much of the public gives engineers credit for improving technology and contributing to economic growth, engineers are seen as having little interest in dealing with people or societal issues (Harris Interactive 2004).

As illustrated in this report, there are compelling reasons why improving public understanding of engineering is a worthy goal for the nation.

THE VALUE OF PUBLIC UNDERSTANDING OF ENGINEERING

Ours is a technology-driven society, and if citizens are to play an informed role in shaping policy decisions that will affect them and their quality of life, they need to have a basic grasp of what technology is, how engineering work is done, and what sorts of considerations and constraints shape the development of various technologies. In other words, they need to be technologically literate (NAE and NRC 2002). The National Academy of Engineering (NAE) publication *Changing the Conversation: Messages for Improving Public Understanding of Engineering* explained it this way:

> A number of important public policy issues, from climate change to the marketing of genetically modified foods, involve scientific and technical issues. Decision making on these and other topics will involve trade-offs, as we attempt to simultaneously manage limited resources while sustaining quality of life. Public discourse and the democratic process could be enhanced if citizens understood more about how engineers are trained and what the practice of engineering entails. (NAE 2008, p. 19)

In addition to helping create citizens who are better able to take part in the democratic decision-making process, technological literacy helps people make better decisions in their own lives. As consumers, individuals must regularly assess the value and usefulness of various technologies—everything from smart phones to electric automobiles—and decide whether they are worth the investment. Understanding something of the engineering process—for example, the role of trade-offs in designing new products—better equips people to make informed choices.

For the United States to maintain its economic competitiveness, it must maintain the capacity for technological innovation (Council on Competitiveness 2004; NAS/NAE/IOM 2007; PCAST 2004). Two of the most important contributors to this capacity are the graduation of qualified engineers from engineering programs (NAE 2005a) and investment in engineering research (NAE 2005b). Both depend in part on the technological literacy of the nation's citizens and policymakers: to the extent that they understand the contributions of engineers and their importance to innovation and competitiveness, the nation is likely to allocate attention and resources to both engineering education and engineering research.

Finally, improved public understanding of engineering should help interest capable students in studying to become engineers. This is true both because it makes engineering more personally appealing to students and because it may positively affect the attitudes toward engineering of individuals, such as family and friends, who influence student decision making about careers (Dietrich and Kracke 2009; Veridian inSight 2009). Student interest in engineering careers, as reflected in expressed intentions to major in engineering in college, reached a 15-year high of 10.3 percent in 2010 (NSF 2012).

A steady supply of qualified, creative engineers is key to industry's ability to innovate, and this, in turn, supports overall economic growth. Attracting students from groups that have historically been significantly underrepresented in the engineering profession is especially key. Women, African Americans, Hispanics, and Native Americans represent a significant reservoir of engineering talent that is not being fully tapped (AAUW 2010; NACME 2008). As the US population becomes more and more diverse, engineering designs should take into account the values and concerns of people across this spectrum, and this aim can be achieved most effectively by having a diverse engineering workforce. For all these reasons it is important to encourage understanding of engineering throughout the US population and not solely in the groups, such as white and Asian males, whose members have long dominated the ranks of engineering.

CHANGING THE CONVERSATION

In 2002 the NAE published *Raising Public Awareness of Engineering*, which reported that although the engineering community had been spending hundreds of millions of dollars each year to advance public understanding of engineering, the messaging seemed to be having little effect. The committee that produced the report concluded that the problem was in large part due to a lack of consistency in the messaging efforts: different segments of the engineering community were offering different messages, with little to no coordination among them, and, more important, the messages had not been developed in any systematic, data-driven way.

In response, the NAE convened the Committee on Public Understanding of Engineering Messages, which had three goals:

1. to identify a small number of messages likely to improve the public understanding of engineering,
2. to test the effectiveness of these messages in a variety of target audiences, and
3. to disseminate the results of the message testing to the engineering community.

The committee worked with a communications firm and a market research company in developing and testing the messages, and the results of the committee's efforts were published as *Changing the Conversation: Messages for Improving Public Understanding of Engineering.*

Positioning Statement, Messages, Taglines, and Survey Results

In deciding on an approach to get people to talk and think differently about engineering, the committee relied on a strategy commonly used by companies and other entities interested in establishing a "brand." It first developed a positioning statement to serve as a guide for the sorts of messages that should be created, and then produced (and tested) a number of messages based on the statement as well as a series of shorter "taglines."

The positioning statement, reproduced in Box 1-1, presents engineering as a field whose practitioners improve people's lives by using their creativity to develop solutions to real-world problems. It intentionally avoids emphasizing the role of mathematics and science in engineering because focus groups and other research had shown that people already understood that engineers use science and math and this message was pushing some students not to consider engineering as a career.

The positioning statement is not intended for public consumption (it is too long and has too many different ideas). Its purpose is to provide guidance to engineers and related organizations on shaping messages about engineering for external audiences. The committee

> **BOX 1-1**
> **Positioning Statement**
>
> No profession unleashes the spirit of innovation like engineering. From research to real-world applications, engineers constantly discover how to improve our lives by creating bold new solutions that connect science to life in unexpected, forward-thinking ways. Few professions turn so many ideas into so many realities. Few have such a direct and positive effect on people's everyday lives. We are counting on engineers and their imaginations to help us meet the needs of the 21st century.

developed five such messages, each articulating an aspect of the engineering brand set forth in the positioning statement. The messages consist of one short sentence that states the message succinctly and a second that expands on it. The committee tested the messages in an online survey of nearly 3,600 adults and teens who were asked how appealing, believable, and personally relevant they found each message. Four of the messages tested well (Box 1-2).[1] To make sure the results would provide statistically valid data for underrepresented groups, the survey population included an oversampling of Hispanics and African Americans.

Adults and teens of both genders rated "Engineers make a world of difference" as the most appealing message. Among teens, boys found this message as appealing as "Engineers are creative problem solvers." The second-favorite message for girls was "Engineering is essential to our health, happiness, and safety" (NAE 2008).

The committee also developed seven taglines (Box 1-3), short phrases that capture some aspect of the positioning statement in a pithy

[1]The fifth message, which was rated least appealing by all groups surveyed, was "Engineers connect science to the real world. They collaborate with scientists and other specialists (such as animators, architects, or chemists) to turn bold new ideas into reality." The project committee speculated the poor result was due to the message's explicit reference to science, and it recommended the message not be used.

BOX 1-2
The Four Messages That Tested Well

- *Engineers make a world of difference.* From new farming equipment and safer drinking water to electric cars and faster microchips, engineers use their knowledge to improve people's lives in meaningful ways.
- *Engineers are creative problem solvers.* They have a vision for how something should work and are dedicated to making it better, faster, or more efficient.
- *Engineers help shape the future.* They use the latest science, tools, and technology to bring ideas to life.
- *Engineering is essential to our health, happiness, and safety.* From the grandest skyscrapers to microscopic medical devices, it is impossible to imagine life without engineering.

SOURCE: NAE (2008).

way that can be used in a magazine advertisement or on a web page or perhaps spoken at the end of a television or radio commercial. Respondents to the online survey were asked to rate these taglines. The most successful among all groups was "Turning dreams into reality," though a greater percentage of boys than girls favored it. "Because dreams need doing" was particularly popular among the teenagers who took the survey and was liked equally well by girls and boys.

Uptake of the Report and Messages

The committee recommended in its report that the engineering community rethink how it presents itself to the general public. In place of traditional outreach efforts, which tended to sell engineering as a remunerative career choice and to emphasize the need for students to be good at science and math, the report called for the community to present engineering as an emotionally satisfying vocation that is inherently creative and concerned with forging a better world. The two pre-

BOX 1-3
Taglines

- Turning ideas into reality
- Because dreams need doing
- Designed to work wonders
- Life takes engineering
- The power to do
- Bolder by design
- Behind the next big thing

SOURCE: NAE (2008).

sentations of engineering are equally true, but the latter is more likely to create the sort of positive image that can help the public understand the value of engineering and attract a larger, more diverse group of students to become engineers.

Changing the Conversation was aimed specifically at "engineering professional societies, technology-intensive industries, colleges of engineering, science and technology centers, and other organizations that communicate with policymakers, K–12 teachers and students, and the public at large about engineering" (NAE 2008, p. viii). The goal was to convince these various organizations to base their own messaging efforts on the new positioning statement, using the specific messages and taglines provided in the report if they wished.

The report did indeed have the desired effect with a number of organizations, and Chapter 2 details many of the efforts to use the messages or related messaging. According to the National Academies Press, which published *Changing the Conversation*, 2,900 copies had been sold as of February 2013. As NAE President Charles Vest noted in 2011, "In the almost three years since the report was published, the notion that the engineering community needs to change its messaging has gained some currency" (Vest 2011, p. 10).

CHALLENGES OF REBRANDING

The challenge of the Changing the Conversation project was to develop a new brand for engineering. As explained in the report,

> By "brand" we mean an association of specific traits in a person's mind that induces behavior. A simple way of understanding this concept might be as a warranty—a promise to perform or deliver. For example, the McDonald's brand promises clean restaurants and food of a known quality. We use this brand as a shortcut in decision making. For example, when traveling on the road, we rely on McDonald's promise to provide a quick, adequate meal. The same thing happens with brands in a grocery or hardware store. As we shop, we make quick judgments based on a brand's promise or warranty. (NAE 2008, p. 24)

The concept of brand can also be applied in areas very different from traditional corporate marketing. Various professions have de facto brands, although they are often not the product of conscious advertising campaigns. Doctors are thought of as "healers," while lawyers generally have a less positive image. The impression that the general public has of engineering is less well defined than that of either doctors or lawyers. It may be somewhat negative (some people think of engineers as "nerdy" or "boring"). The public strongly associates engineering with high ability in mathematics and science, an image often promoted by engineers themselves. But research has shown that the biggest problem is simply that most people have only a vague idea of what engineers actually do (NAE 2008).

It is possible for a profession to rebrand itself, to create a new and more favorable image and understanding of itself in the mind of the public. This has been attempted, for example, by actuaries (Beuerlein 2006), accountants (AICPA 2012), and, in the United Kingdom, the nursing profession (www.nursingthefuture.org.uk). But rebranding can be difficult, time-consuming, and expensive, and the engineering profession faces a number of particular challenges (Baranowski 2011).

The first challenge is dealing with the existing brand. The stereotype of engineers is due in part to messaging efforts of the engineering community itself, which has often emphasized the role of science and mathematics in engineering and failed to discuss how engineers grapple with real-world problems, including societal and environmental issues. It should be possible to move away from this brand, but it will require time and a consistent and unified messaging effort from the engineering community.

A second challenge will be to achieve coordination and consistency in the rebranding effort. There is no single voice that speaks for the engineering community and presents an image of engineering to the broader world. Each sector of the community has a stake in the public perception of the field, but each also has unique goals, capabilities, and parochial interests. This poses obvious challenges to the coordinated, consistent delivery of messages.

A third challenge centers on language and what words mean to different people. For example, many people do not distinguish the word "engineering" from science or technology. And engineers themselves do not always agree on what engineers do or how to explain it to the public. This makes it difficult to get a clear brand across. The issue is further complicated when those in the community attempt to describe the differences between subdisciplines of engineering, such as mechanical, electrical, or civil.

Finally, a rebranding program must deal with the "promise gap"—the gap between what is described in the rebranding effort versus what people encounter in real life. For instance, if students decide to major in engineering because they have heard that it is a creative profession that enables people to help address societal problems but then experience several years of classes focused mostly on solving science and math problems, they may feel misled and may convey their disillusionment to others.

CHANGING THE CONVERSATION: FROM RESEARCH TO ACTION

In recognition of the challenge of rebranding engineering, in 2010 the NAE, with funding from the National Science Foundation, initiated a follow-up project called Changing the Conversation: From Research to Action. This project was intended to provide encouragement and assistance to the engineering community in adopting and acting on the positioning statement, messages, and taglines developed in the earlier effort. The statement of task for the project outlined three goals:

1. sponsor a high-level stakeholders' workshop to develop support for a coordinated national messaging campaign,
2. produce an online messaging resource (a "toolkit") for use by the engineering community, and
3. publish an "action plan" containing strategic and tactical recommendations for the engineering community to effectively promote a more positive and accurate image of engineering.

The project was overseen by a 10-member advisory committee co-chaired by Ellen Kullman, chair of the board and CEO of DuPont, and NAE President Charles M. Vest (see biographies of committee members in Appendix A).

At the end of 2010, the committee hosted a 2-day workshop with several dozen high-level decision makers representing key stakeholder groups in the engineering community. The workshop was designed to educate the attendees about the original Changing the Conversation project; to discuss how the project's messages had been put to use by selected organizations; and to identify further actions the engineering community might take to more fully implement the CTC positioning statement, messages, and taglines, including

- using the messages and taglines in new outreach activities;
- disseminating new messaging consistent with *Changing the Conversation* through specific communications tactics and tools, such as web-based media;
- and supporting a coordinated messaging effort, including possibly a national-scale advertising campaign.

The third goal, an action plan, is the subject of this report, which presents a series of steps that can—and, the committee believes, should—be taken to encourage the new ways of thinking about engineering described in *Changing the Conversation.*

OUTLINE OF THIS REPORT

Following this introductory chapter, Chapter 2 examines what has been accomplished since the release of *Changing the Conversation.* It provides evidence of improved messaging from the engineering community, including both messaging that is based directly on messages and taglines from *Changing the Conversation* and messaging that is in the spirit of that report but does not use the report's messages and taglines explicitly. The chapter also describes the creation of several new tools and resources for members of the engineering community interested in taking part in the rebranding effort. These include the Changing the Conversation website, a Changing the Conversation Facebook page, and a train-the-trainer effort developed by the National Engineers Week Foundation (NEWF).

Chapter 3 lays out the details of an action plan, suggesting specific steps for various components of the engineering community: corporations, professional societies, government agencies, engineering schools, science and technology centers, and the NAE. We discuss what each can hope to gain by changing the conversations, actions already taken, constraints, and ways to move forward. We also make the case for coordination among these components and suggest ways to achieve such coordination. The chapter ends with a set of recommendations.

AUDIENCES FOR THIS REPORT

This report is aimed at multiple audiences in the engineering community, primarily those that are already engaged in some sort of messaging or outreach efforts, but also those that might be inspired to join the rebranding campaign called for in *Changing the Conversation.* The professional engineering societies, such as the American Society for Engineering Education, Institute of Electrical and Electronics Engineers, ASME, American Society of Civil Engineers, and Society of Women Engineers compose one key audience. These societies generally have outreach efforts, and their members take part in those efforts, such as by volunteering in schools or other venues. In addition, there are organizations whose main function is outreach, such as NEWF, and engineering-related contests run by organizations such as US FIRST/FIRST Lego League, BEST, and the Toshiba Exploravision awards.

In the business world, a variety of corporate and industry associations are interested in engineering; like the professional societies, these organizations generally have outreach programs that could be modified to convey the messages presented in *Changing the Conversation.* Furthermore, numerous companies depend heavily on engineering and technical talent and many of them have outreach efforts, often through corporate foundation education units or corporate marketing departments.

There are over 300 undergraduate engineering programs in the United States, which typically have outreach materials for prospective students. And because deans and faculty members have a great deal of influence over how students view engineering, they are part of the intended audience for this report. There are also a number of associations connected with education in STEM, such as the National Science Teachers Association, National Council of Teachers of Mathematics, and International Technology and Engineering Educators Association. Each has activities aimed at informing public views of engineering. Because of their role in advising about career options, high school guidance counselors, through organizations such as the National Association for College Admission Counseling, also may benefit by being better informed about engineering.

Governmental audience members include the national laboratories and other federal R&D labs as well as government agencies that deal with engineering, such as the National Science Foundation, National Institute of Standards and Technology, National Aeronautics and Space Administration, Department of Defense, and Department of Energy.

There are many science and technology museums around the country—the Exploratorium in San Francisco, the Tech Museum of San Jose, the Boston Museum of Science, the Oregon Museum of Science and Industry, and many more. Their purpose is to explain science and technology to the public, so they are natural targets for the suggestions presented in both *Changing the Conversation* and this report.

Finally, the NAE itself, through its projects and the influence and visibility of its leadership and members, is well positioned to influence public understanding of engineering.

Changing how the general public thinks of engineering will require the participation of all these segments and representatives of the engineering community, all pulling in the same direction. This report offers suggestions to promote that participation in support of accomplishing the goal of "rebranding" engineering for the public.

REFERENCES

AAUW (American Association of University Women). 2010. Why So Few? Women in Science, Technology, Engineering, and Mathematics. Available online at www.aauw.org/learn/research/upload/whysofew.pdf (accessed January 31, 2013).

AICPA (American Institute of Certified Public Accountants). 2012. Start Here, Go Places. Available online at www.startheregoplaces.com (accessed June 10, 2012).

Baranowski, M. 2011. Rebranding engineering: Challenges and opportunities. *The Bridge* 41(2): 12–16. Available online at www.nae.edu/Publications/Bridge/51063/51088.aspx (accessed June 20, 2012).

Beuerlein, R. 2006. One voice, one brand. Presidential address. *The Actuary Magazine.* April/May. Available online at www.soa.org/library/newsletters/the-actuary-magazine/2006/april/act-pres-address.aspx (accessed June 10, 2012).

Council on Competitiveness. 2004. *Innovate America: Thriving in a World of Challenge and Change.* Available online at www.compete.org/images/uploads/File/PDF%20Files/NII_Innovate_America.pdf (accessed June 7, 2012).

Cunningham, C.M., C.P. Lachapelle, and A. Lindgren-Streicher. 2006. Elementary teachers' understandings of engineering and technology. In *American Society for Engineering Education Annual Conference & Exposition.* Chicago: American Society for Engineering Education.

Dietrich, J. and B. Kracke. 2009. Career-specific parental behaviors in adolescents' development. *Journal of Vocational Behavior* 75(2): 109–119.

Harris Interactive. 2004. American Perspectives on Engineers and Engineering. Poll conducted for the American Association of Engineering Societies. Final report, February 13, 2004. Available online at www.aaes.org/harris_2004_files/frame.htm (accessed June 13, 2012).

Lachapelle, C.P., P. Preeva, J. Hertel, and C.M. Cunningham. 2012. What is engineering? A survey of elementary students. Paper presented at the 2nd P–12 Engineering and Design Education Research Summit, Washington, DC.

NACME (National Action Council for Minorities in Engineering). 2008. *Confronting the "New" American Dilemma—NACME 2008 Data Book.* Available online at www.nacme.org/user/docs/DataBook.pdf (accessed January 31, 2013).

NAE (National Academy of Engineering). 2002. *Raising Public Awareness of Engineering.* Washington: National Academies Press.

NAE. 2005a. *Educating the Engineer of 2020: Adapting Engineering to the New Century.* Washington: National Academies Press.

NAE. 2005b. *Engineering Research and America's Future: Meeting the Challenges of a Global Economy.* Washington: National Academies Press.

NAE. 2008. *Changing the Conversation: Messages for Improving Public Understanding of Engineering.* Washington: National Academies Press.

NAE and NRC (National Research Council). 2002. *Technically Speaking: Why All Americans Need to Know More About Technology.* Washington: National Academies Press.

NAS/NAE/IOM (National Academy of Sciences, NAE, and Institute of Medicine). 2007. *Rising Above the Gathering Storm: Energizing and Employing America for a Brighter Economic Future.* Washington: National Academies Press.

NSF (National Science Foundation). 2012. Science and Engineering Indicators. Appendix Table 2-12, Freshmen intending S&E major, by field, sex, and race/ethnicity: 1995–2010. Available online at www.nsf.gov/statistics/seind12/append/c2/at02-12.xls (accessed February 15, 2013).

PCAST (President's Council of Advisors on Science and Technology). 2004. *Sustaining the Nation's Innovation Ecosystem: Maintaining the Strength of Our Science and Engineering Capabilities.* Available online at www.calstate.edu/psm/docs/1-11-07PSM_Jun2004.pdf (accessed June 7, 2012).

Veridian inSight. 2009. *"Engineer Your Life" Evaluation Report for Year 2: Executive Summary.* Prepared by Veridian inSight, LLC, with contributions from American Institutes for Research. Unpublished.

Vest, C.M. 2011. The image problem for engineering: An overview. *The Bridge* 41(2): 5–11.

2
WHAT HAS BEEN ACCOMPLISHED?

Since the release in 2008 of *Changing the Conversation*, a number of institutions have either directly used or adapted its messages and taglines. Other organizations, mostly in industry, have created messaging about engineering that, while not based specifically on the CTC report, is very much in the spirit of its recommendation to portray engineering as a profession that makes a difference in people's lives. The National Academy of Engineering (NAE), for its part, has undertaken new projects that build on the report, including (1) development of an online "toolkit" (www.engineeringmessages.org) that supports those engaged in messaging efforts and (2) creation of a Facebook page (www.facebook.com/engineersctc) to facilitate broader discussion of engineering messaging through social media.

MESSAGING BASED ON *CHANGING THE CONVERSATION*

We present several cases illustrating the implementation and impacts of messaging based directly on the CTC positioning statement, messages, and taglines. The cases represent some of the most ambitious and well-documented attempts to date to reposition engineering.

Society of Women Engineers

The Society of Women Engineers (SWE), established in 1950, has long worked to improve the image of engineering, particularly among females and underrepresented minorities. Its outreach efforts have included print and web-based materials for the public, society-sponsored programs, and materials and training for the more than 400 SWE sections. With the publication of *Changing the Conversation*, SWE reworked all of its messaging products to align with the CTC positioning statement and messages.

The SWE's rebranding efforts took place on a number of fronts, not only in its print and web-based outreach materials but also at its signature events, *Invent It. Build It.* and *Wow! That's Engineering!*, both designed to increase interest in engineering among girls. These events include sessions for the teachers, parents, and other adults who accompany the girls so that these adults can reinforce the CTC messages with the girls. And because individual SWE sections use materials from these events in their outreach efforts, the CTC messages get an even broader exposure. In addition, SWE operates an online training center for volunteers who wish to increase engineering's appeal among girls. More than 1,600 volunteers have completed this "basic training," which provides details about the CTC messages and their use.

SWE also collaborates with three other engineering professional organizations—the National Society of Black Engineers, Society of Hispanic Professional Engineers, and American Indian Science and Engineering Society—to increase minority participation in engineering. The collaboration, referred to as 4Δ (www.outreach4change.org), is now using *Changing the Conversation* as the basis for its own messages. The test populations that reacted to the original CTC messages included significant numbers of black and Hispanic adults and teens, so members of the collaboration are confident that the messages will be effective with these audiences.

Further extending its outreach and impact, SWE works with engineering companies through its Corporate Partnership Council, whose 62 members are leading employers of engineers. After a presentation on the CTC messages, a number of the participating companies incorporated the messages in their corporate outreach. For example, Dell modified its poster for National Engineers Week (E-Week; www.eweek.org), the long-standing celebration of engineering held each year in February, to use CTC messages. Abbott Labs, which offers a career fair for middle and high school students, gave the CTC tagline "Turning dreams into reality" a prominent place at the fair, and at a high school program company engineers selected a CTC tagline and connected it with the type of work they do. GM incorporated the CTC messaging in a K–12 engineering curriculum project, A World in Motion®, that it sponsors in Detroit. And at Agilent, the newsletter of a women's affinity group described the CTC messaging to its readers.

SWE members discussed CTC messaging at a session of the 2011 Annual Conference for Women Engineers. At the end of the session, participants were asked to take specific steps to promote the messages when they returned to their home institutions. The messages and taglines were also visible in a number of media at the conference, from the inclusion of "Engineers make a world of difference" on slides shown at the awards banquet to the printing of a CTC tagline on the shirts of the Genentech recruiting team.

University of Colorado–Boulder

At the University of Colorado–Boulder, the College of Engineering undertook a major rebranding effort based on *Changing the Conversation.* Jackie Sullivan, the college's associate dean for inclusive excellence, served on the CTC committee and took the lead in introducing CTC messages on campus (Sullivan 2010). She began by talking with the engineering college's communications staff; once they were convinced of the value of the CTC approach, a number of campuswide workshops were held to generate ideas for applying the approach at CU-Boulder. Brainstorming at these workshops led to the design by the engineering communications staff of new engineering recruiting brochures with aspirational messages. Recruiting postcards now feature CTC messages and ask prospective students "How will the world CU?"

Soon individual engineering departments embraced and adapted the enhanced recruiting efforts. The Department of Electrical and Computer Engineering created a postcard with CTC messages to entice high school students to attend its design expo (Figure 2-1), and the Department of Mechanical Engineering produced a brochure based on CTC messages to appeal to female and minority students. At about the same time, the College of Engineering produced new recruiting postcards with the tagline "Engineer your life"[1] and pictures showing diverse college students enjoying a variety of extracurricular activities: flying a plane, riding a bicycle, and salsa dancing.

Two CTC messages and a tagline are used pervasively at the college: "Engineers make a world of difference" and "Engineers help shape the future" resonate with the campus community's green culture and renewable energy focus, and "Because dreams need doing" is popular because of its strong aspirational tone (research in the original CTC project also found this message to be equally appealing to young men and women).

In 2012 the number of minority students who enrolled in the College of Engineering, 109, was more than double the average number enrolled annually between 2004 and 2008. Over the past 4 years, the 6-year retention (through to graduation) rate of minority students increased from 34 percent to 53 percent, a 56 percent jump. Annual average female first-year enrollment, which between 2004 and 2008 stood at 137, rose to an average of 178 women per year for the 2009–2012 period. During the 2004–2008 span, women constituted 20 percent of the engineering class, while during the more recent period, their share increased to 25 percent, a 25 percent rise. It is impossible to prove that these increases were due to the new messages, but they are certainly encouraging.

Engineer Your Life Website

The Engineer Your Life (EYL) website is the centerpiece of a national campaign to encourage college-bound girls to explore engineering.

---

[1] "Engineer your life" is a tagline, the name of a website (www.engineeryourlife.org), and a project to interest high school girls in the possibilities offered by a career in engineering. The tagline was developed with the help of the same PR firm (BBMG) and market research company (Global Strategy Group) that helped create the CTC messages and taglines. More on the website in the next section.

FIGURE 2-1 Flyer for the 2009 CU–Boulder Electrical, Computer, and Energy Engineering (ECEE) Design Expo

The campaign, funded in large part by the National Science Foundation, is overseen by a partnership of WGBH Boston, the NAE, and the Extraordinary Women Engineers Project coalition. The EYL messages are spread by more than 100 coalition members, which include colleges and universities, engineering and educational organizations, and corporations such as 3M, DuPont, Intel, and Lockheed Martin. Besides reaching out to girls through the EYL website and extensive social networking, the project offers a variety of resources for communicating with young people about engineering careers for use by parents, school counselors, engineering professionals, and college recruiters.

In addition to the CTC messages, the EYL project created new ones similar in feel and approach but developed and tested specifically with girls' career aspirations and motivations in mind. The four best-testing messages, according to EYL survey results (Wolsky 2011, p. 32), were the following:

- **Live your life, love what you do.** Engineering will challenge you to turn dreams into realities while giving you the chance to travel, work with inspiring people, and give back to your community.
- **Creativity has its rewards.** Women engineers are respected, recognized, and financially rewarded for their innovative thinking and creative solutions.
- **Make a world of difference.** From small villages to big cities, organic farms to mountaintops, deep-sea labs to outer space, women engineers are going where there is the greatest need and making a lasting contribution.
- **Explore possibilities.** Women engineers often use their skills to go into business, medicine, law, or government. An engineering education will prepare you for many different careers.

The EYL website features video profiles of young women engineers, descriptions of interesting engineering jobs, and a list of "Ten great reasons why you'll love" engineering (Figure 2-2). EYL's social networking channels offer posts and discussions on six Facebook fan pages: *EYL, Creativity, Making a Difference, Dream Jobs, Design,* and

FIGURE 2-2 Screenshot of Engineer Your Life "Why Engineering?" Page Featuring Excerpts from Several CTC Messages

The Engineer's Pledge (Box 2-1). EYL also provides training sessions to ensure that thousands of engineers and educators know how to talk about engineering positively, and effectively, to girls.

IEEE Video Contest

The US organization of the Institute of Electrical and Electronics Engineers (IEEE-USA) took a very different approach to spreading the messages of *Changing the Conversation.* During the 2008–2009 school year, it launched an annual scholarship contest for undergraduate students to create a 90-second video promoting engineering to 11- to 13-year-olds, based on the CTC theme "Engineers make a world of difference" (Figure 2-3). There are four categories of prizes totaling $5,000.

Initially the competition was open to teams composed solely of IEEE student members, but teams now can include non-IEEE members as long as there is at least one IEEE student member participating. The

BOX 2-1
The Engineer's Pledge

The Engineer's Pledge appears on the Engineer Your Life Facebook page and the EYL project website. It asks visitors to help give engineering a better image by letting people know how exciting and rewarding an engineering career can be. Engineers need to change the way they talk about engineering, it says. By liking the Facebook page, visitors are pledging to

- tell people about the creative aspects of engineering,
- promote the collaborative nature of engineering, and
- talk about how engineering makes a difference.

As of early 2013, nearly 1,800 people had liked the page.

SOURCE: www.facebook.com/engineerpledge

FIGURE 2-3 IEEE-USA Notice Describing Its 2011–2012 "World of Difference" Video Contest

contest is advertised on the IEEE-USA and IEEE student websites, and announcements are distributed through IEEE student email lists, ads in the IEEE student magazine, and postings on Facebook.

The winning entries are announced during each year's E-Week and are showcased in various ways, including through posting on the Design Squad website and YouTube; a prizewinning entry in the 2011–2012 contest created by an Ohio University student was viewed more than 4,000 times on YouTube. The winning videos are also provided on DVDs to every IEEE student chapter in the United States.

Although there have been relatively few entries in the contest each year—in part because few undergraduates outside of IEEE student chapters know about the contest—the quality of the entries has improved consistently, and the contest is an excellent example of a creative approach to spreading the messages articulated in *Changing the Conversation.* The winning entries are selected by a panel of judges comprising Nate Ball, a mechanical engineer and cohost of PBS's Design Squad Nation, and two graduate student members of IEEE.

National Engineers Week Foundation

National Engineers Week Foundation (NEWF) has been applying the CTC messaging for years. In fact, even before the 2008 publication of *Changing the Conversation,* NEWF used one of the best-testing taglines from the CTC project, "Turning ideas into reality," as well as the message "Engineers make a world of difference" in posters and other promotional materials. The theme for the 2012 event drew on a different tagline: "Seven billion people. Seven billion dreams. Seven billion chances for engineers to turn dreams into reality" (Figure 2-4). For its 2010–2011 Future City competition, NEWF used the tagline "Dreams need doing" in the promotional poster; teacher handbook; volunteer handbook; and handouts for sponsors, parents, and volunteers. E-Week also uses the CTC messages when promoting its Family Engineering Day. Held each year during National Engineers Week, the event draws between 6,000 and 10,000 children and adults to the National Building Museum in Washington, DC, making it the largest public event the facility hosts. And NEWF has used the CTC messaging in ads in

FIGURE 2-4 2012 National Engineers Week Promotion Poster

USA Today announcing its New Faces in Engineering program, which recognizes the work of early-career engineers.

National Academy of Engineering

The NAE has spread the CTC messages and taglines in a variety of ways, in part by linking them with its programs and other activities. For example, to spread the word about the Grand Challenges for Engineering project, which identified 14 important issues facing humanity that will require engineering talent to address, the Academy created a bumper sticker (Figure 2-5) that combines the project's URL (www.engineeringchallenges.org) with the tagline "Because dreams need doing." Some 11,000 of the stickers have been distributed.

In addition, the Grand Challenges website invites visitors to "See how engineers can make a world of difference." A popup window links to four specially produced videos on the main challenge themes: sus-

FIGURE 2-5 Grand Challenges for Engineering Bumper Sticker

tainability, health, security, and joy of living (www.nae.edu/Activities/Projects/grand-challenges-project/57302.aspx). A brief fifth video, *Build Your Dream,* synthesizes the themes as a narrator says, "Out of each generation step a few bold individuals—bright dreamers and tenacious doers—who envision a smarter world, who evolve the promise of today into the joy of tomorrow. We call them engineers." The Grand Challenges site also provides a link to the CTC homepage.

The NAE used the message "Engineers Turn Ideas into Reality" in a poster to solicit nominations for the 2014 Bernard M. Gordon Prize, a $500,000 award that recognizes innovations in engineering and technology education (Figure 2-6). Some 500 copies of the poster were distributed to engineering schools and engineering-related organizations.

As part of a National Academies–wide effort, in 2010 the NAE participated in the inaugural Science and Engineering Festival, an event on the National Mall in Washington that drew a half-million people to 550 hands-on activities and exhibits with a science or engineering focus. The NAE exhibit, developed in partnership with Walt Disney Studios, focused on the technology and engineering featured in the movie *Tron: Legacy* and also made connections to the Grand Challenges for Engineering. Signage for the entire National Academies effort used the CTC message "Because dreams need doing" (Figure 2-7).

The NAE devoted the summer 2011 issue of its quarterly magazine *The Bridge* to the subject of engineering messaging. The issue featured an introductory editorial by DuPont Chair of the Board and CEO Ellen Kullman, an article detailing engineering's "image problem" by NAE President Charles M. Vest, and four additional articles addressing topics such as how rebranding initiatives are designed, the elements of

FIGURE 2-6 2014 Gordon Prize Poster

FIGURE 2-7 National Academies Signage for 2010 USA Science and Engineering Festival

large-scale outreach campaigns, and use of the CTC messages at the University of Colorado–Boulder to recruit new students (described above). *The Bridge* is mailed to about 7,000 subscribers, including all members of Congress, NAE members, and participants in the NAE's Frontiers of Engineering program (www.naefrontiers.org).

Dr. Vest has actively promoted and encouraged the use of the CTC messages. For example, he noted their importance in his speech on the future of engineering at the 2011 NAE Annual Meeting (Vest 2011) and in April 2012 when he spoke to representatives of some 60 engineering professional groups at the NAE Convocation of Professional Engineering Societies (Vest 2012). Ms. Kullman promoted the CTC messages in speeches to the Philadelphia Chapter of the American Institute of Chemical Engineers in April 2010 and the executive committee of the Society of Chemical Industry Americas Group in March 2011. In 2012 she encouraged use of the messages during a jobs and competitiveness discussion at the Council on Foreign Relations, a panel discussion at the U.S. News STEM Solutions conference, an NBC Education Nation roundtable, and an award acceptance speech at the National Action Council for Minorities in Engineering.

MESSAGING IN THE SPIRIT OF *CHANGING THE CONVERSATION*

A number of noteworthy messaging efforts both before and since the publication of *Changing the Conversation* are very much in the spirit of the CTC messages and taglines. These efforts are of two types: (1) those that have customized or adapted the CTC messages to an organization's needs or goals and (2) those that have developed independently of the CTC project but that are consistent in tone and meaning with its messages. The latter not only provide additional examples of effective messaging but also suggest movement in the engineering community toward messages that portray engineering as an inspiring career option for people to take on interesting challenges and help make the world a better place.

Customized or Adapted CTC Messages

The American Council of Engineering Companies of New York (ACEC New York) began reworking its advertising and public relations efforts in 2006. The group had traditionally been concerned mainly with influencing legislative decisions that affected its member firms' interests, but two developments led to a reconsideration. First, the results of a survey of its members showed that many engineering managers were worried about the supply of new engineers. And second, members had learned about the NAE's work on public understanding of engineering and its call for a new approach to messaging emphasizing the creativity of the profession and the potential to make a difference through a career in engineering.

Spurred by these developments, ACEC New York undertook a "vision campaign" aimed at conveying the creative nature of engineering in order to inspire young people to consider it as a profession. With a budget of less than $40,000, the organization had to look for ways to grab attention quickly with a universal message based on the foundation of NAE research and market-tested messages. The result was a compelling visual image combined with messages and taglines—"Your vision can change the world" and "Engineering . . . takes creativity, imagination, and vision"—that capture the excitement and importance of engineering. A young person was pictured looking through a pair of binoculars directly at the viewer, with two different images in the two glass lenses of the binoculars: a cityscape and a view of the earth from many miles above the surface (Figure 2-8).

During the 2009 E-Week, the ad achieved the ultimate in visibility. With funding from the Siemens Corporation, ACEC New York

FIGURE 2-8 ACEC New York Ad Promoting Engineering

transformed the static ad into an animated 30-second spot that was displayed on the ABC Studios Jumbotron screen in New York's Times Square. It ran about once an hour throughout the week, reaching an estimated 1.2 million viewers (O'Grady 2010).

Independently Developed Messages

A number of organizations, mostly in industry, have included and sometimes highlighted engineers and engineering in their marketing efforts through either traditional corporate advertising (i.e., touting the company's products or brand) or outreach to attract new employees.

A good example of the former is a Lenovo online video that promotes the company's laptop computers (Patterson 2010). The computers are shown being tested in various extreme conditions—under water, in the middle of the desert, in outer space—with the implication that the laptops are completely reliable. But the ad's tagline—"From the world's best engineers come the world's best-engineered PCs"—makes the role of engineers explicit, and the images throughout the video feature engineers working in very interesting, even exotic, settings.

A good example of recruiting-focused messaging is from the Norwegian company Norsk Hydro, which made a series of amusing commercials showing teenagers using technology to pull off creative pranks. In one, a group of youths modifies a train track to create a giant loop-the-loop roller coaster: a train comes zooming down the track, zips up and around the vertical loop, and then continues speeding down the track as the kids cheer (Norsk Hydro 2008a). In another, a group of teenagers rig an adult's car so that they can steer it using a radio-control device; when the adult gets in, they send the car careening around the neighborhood, crashing through a garage, and driving tilted up on the driver's-side wheels (Norsk Hydro 2008b). At the end of each mini-story flashes the message "There are many young engineers. We can't wait till they grow up."

Similarly, Texas Instruments created a series of "Thank an engineer" videos that offer humorous looks at how various technologies add to people's lives by, for example, envisioning a world without a particular technology. A piece on MP3 players shows people getting their music from cassette tape–playing boom boxes that they carry with them as they go jogging, ride a bicycle, or mow the lawn (Texas Instruments 2011). The video ends by saying, "TI salutes the engineers who have enhanced the world in so many ways. Thank you for all you do."

In a more serious vein, a 6-minute Cisco (2007) recruiting video features a group of women engineers talking about what they do for the computer networking company. The images are chosen to send the message that these are just "regular" people—they are shown skateboarding, playing field hockey, holding a kitten—and the women directly address the point that they had been raised to think that engineering is just for boys but discovered that it isn't. The women's descriptions of what they do convey both the creativity of their jobs and the satisfaction they get from doing things to make people's lives better. One woman comments, "I could have been a doctor or . . . a teacher, but I chose to be an engineer, and I still get to help people."

A number of other companies have produced advertising or recruiting materials that reflect the spirit of *Changing the Conversation* (Table 2-1). They focus on the value of engineering to people's lives and on the creativity of the engineering profession, and show that a career in engineering is within reach for many young people who have vision and a desire to solve problems and help people.

TABLE 2-1 Additional Examples of Industry Messaging in the Spirit of *Changing the Conversation*

Sponsor company	Summary	Message(s)	Link
ABB	"Future Astronauts" is a 30-second ad showing a diverse group of kids around a conference table imagining what is needed to put humans on Mars.	"We'll need lots of engineers" and "Ingenuity at Work."	www.youtube.com/watch?v=P0j87GdB3to
ExxonMobil	"ExxonMobil Math and Science" is a video featuring two company engineers who talk about being interested in engineering at a young age. Math and science skills are emphasized.	"ExxonMobil Math and Science" is described as a program that helps harness young students' interest in "tinkering" early so that they can grow up to "solve the world's next great challenges."	www.youtube.com/watch?v=1ReLtjvALN4
Lockheed Martin	"How" is a series of 30-second ads and longer videos showing how Lockheed Martin creates innovative products and services to solve problems.	"'How' is the word that makes all the difference."	www.lockheedmartin.com/us/how/index.html
GE	The "What Works Project" produced this Super Bowl commercial that explains how GE power turbines help make cold beer.	The tagline is "GE Works," and one of the GE employees says, "It's nice to know what you're building is going to do something for the world."	www.youtube.com/watch?v=DgNStjaxOCM&feature=player_embedded

CREATION OF NEW TOOLS AND RESOURCES

The NAE has developed tools and resources to help spread the CTC messages. Other organizations have provided assistance as well, with the result that it is now much easier for interested parties to join in the effort to change the conversation about engineering.

The CTC Website

The development of a positioning statement, messages, and taglines was just the first step in changing the conversation about engineering. Next it was necessary to implement and disseminate them, and the first major tool created for that task was the CTC website (www.engineeringmessages.org), designed to incorporate the artwork and speech-balloon imagery from the 2008 report (Figure 2-9). The site was developed by the committee in concert with Diamax, a digital design and interactive strategy firm that has extensive experience with engineering. A beta version of the site was shared with various audiences, including the National Engineers Week Steering Committee, prior to the site's launch in early 2011. The committee used both email and face-to-face meetings at conferences to notify stakeholders, such as engineering schools and engineering professional societies, about the site and to encourage them to link to it. NAE President Charles Vest sent a note to all NAE members informing them of the site's launch.

The site lays out the problem and presents the solution proposed in *Changing the Conversation.* Under "The Problem" is information about the poor public understanding of engineering along with the reasons

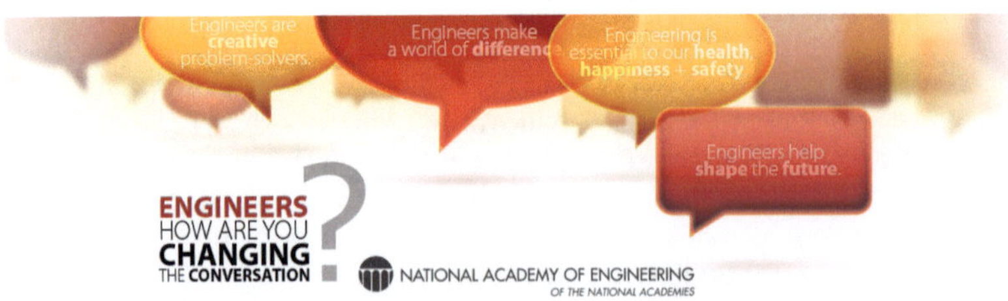

FIGURE 2-9 Screenshot of CTC Homepage

for and consequences of that poor understanding. This section explains how engineers fail to present engineering in a compelling or exciting way for the public. And it notes the lack of diversity among engineers, with comments on the potential effects on the nation's economic competitiveness.

Under "The Solution," visitors can read the complete *Changing the Conversation* report online and then go to a page that provides all the recommended messages and taglines in one place. In addition, the site offers tips for effectively using the messages, with recommendations such as "Show, don't tell" and "Use multiple media." The site also features nearly 140 examples of engineering messaging efforts—videos, web pages, brochures, posters, and magazine covers that use the CTC messages and taglines themselves and those in the spirit of CTC.

Finally, the CTC website offers resources to help people take effective action. There are, for instance, case studies of organizations that have used the CTC messages in creative ways. Visitors to the site can describe other actions that they know about and post links to sites with more information on those actions. In addition to a CTC blog, where NAE staff provide relevant information quickly and informally, there are downloadable and customizable CTC posters, bookmarks, and door hangers. And there is a directory of individuals interested and active in changing the conversation titled *Who's Changing the Conversation?* To encourage sharing via social media, nearly every page has tabs for Google+, Facebook, Twitter, and Pinterest.

The goal is for the CTC website to become both the central meeting place for all who wish to change the public perception of engineering and a primary source of information on the best ways to achieve that transformation. By the end of 2012, the website had about 420 registered members, and over the course of the year it received nearly 50,000 visits. According to Google Analytics, since the site was launched, about 9 percent of visitors came to the site through CTC links appearing on other websites (Box 2-2).

BOX 2-2
Sample of Organizations Linking to www.engineeringmessages.org

White House (Champions of Change Blog)
Center for Engineering Education Research, Michigan State University
University of Colorado–Boulder School of Engineering
Santa Clara University School of Engineering
University of Virginia Climate Change Partnership
Purdue University School of Engineering
Pennsylvania State University Engineering Ambassadors Program
Rensselaer Engineering Ambassadors Program
American Institute of Mining, Metallurgical and Petroleum Engineers
American Society of Civil Engineers
American Society for Engineering Education
Institute for Electrical and Electronics Engineers
National Action Council for Minorities in Engineering
Tau Beta Pi (The Engineering Honor Society)
International Technology and Engineering Educators Association
United Engineering Foundation
National Defense Industrial Association
STEM Equity Pipeline
STEM Connector
TeachEngineering
King Features (Hearst Corporation)
The Science and Entertainment Exchange
Engineer Your Life
Grand Challenges for Engineering

NOTE: Links appear on homepage or subpage. List accurate as of February 2013.

The CTC Facebook Site

Facebook has become the most important social networking site in the world. With more than 1 billion active users, it is a place not only where individuals connect and interact with other individuals but also where businesses and organizations maintain a presence in order to increase their visibility and manage their images. IBM and Intel have Facebook pages, for example, as do GE, DuPont, Merck, and most other major corporations. IEEE and many other professional societies also maintain Facebook sites. Such a presence is increasingly important as people spend more time each day on Facebook and obtain an increasingly large amount of information from its social network. Thus it made sense for the NAE to establish a Facebook site to encourage awareness and promote discussion about *Changing the Conversation.* Created and supported with funding from the United Engineering Foundation (www.uefoundation.org; UEF), the site was "liked" by over 5,300 people within 11 months of its launch.

Called "Engineers Changing the Conversation," the CTC Facebook page uses graphics similar to those of the CTC website. Visitors to the site can find information about various engineering events and watch engineering-related videos. The major component of the site, however, is the collection of postings about improved messaging and about engineering in general, often with links for visitors interested in more information.

There is, for example, a note concerning a *Forbes* article written by NAE member Andrew Viterbi, cofounder of Qualcomm, arguing for the importance of getting young people interested in engineering, along with a link to the article. Another post describes a teacher's 17-page discussion guide for a class that "deliberately focuses on engineering concepts and outcomes besides math and science," again with a link to the relevant material. A third post incorporates two CTC phrases—"changing the conversation" and "turning dreams into reality"—in reference to a Northrop Grumman program for high school students interested in engineering. Several entries describe engineers who might be inspiring to young people, such as a 22-year-old recent graduate who is interested in sustainable living and a female engineer from Queensland who talks about "making a difference in the community." The mission of The Works, a museum in Minnesota, is to "awaken every child's inner engineer." Many posts are followed by comments that sometimes become a back-and-forth conversation.

Facebook offers a way to reach out to young people that was not available 10 years ago. We hope the CTC site will both support development of a community of people who are interested in changing the conversation about engineering and provide a forum for the exchange of ideas, advice, and encouragement.

E-Week Train-the-Trainer Effort

As has been clear since before the publication of *Changing the Conversation*, propagating a new image of engineering will require not only websites and marketing campaigns but also the work of many individual proponents who spread the message in one-on-one and group interactions. With this in mind, the National Engineers Week Foundation (NEWF), which sponsors E-Week and other activities, developed a CTC train-the-trainer program for outreach volunteers. The program, which is also supported with funds from UEF, is intended to promote implementation of the CTC messages and taglines.

First offered in 2012, the training program consisted of a PowerPoint presentation summarizing the problems that led to the CTC rebranding effort; explaining the research behind the selection of the program's positioning statement, messages, and taglines; describing the messages and taglines; and providing tips and examples for using them. NEWF pilot tested the PowerPoint in face-to-face sessions at a coordinators' retreat for the Future City program (www.futurecity.org), an annual engineering-focused competition for middle school students organized by NEWF, and at a Northrop Grumman Women's conference. The training was further refined via virtual (webinar-based) sessions conducted for members of the National Society of Professional Engineers, the NEWF's steering committee and diversity council, "fans" of the NAE Facebook page, and the NAE membership. The final PowerPoint presentation and links to the digital files from the webinars were then posted on the CTC website.

Ideas from the training were incorporated in a two-page CTC messaging "tip sheet" included in packets mailed to all entrants in the 2013 Future City competition. The tip sheet is also available on the CTC website.

REFERENCES

Cisco. 2007. So you want to be an engineer? Available online at www.youtube.com/watch?v=usXRaL8cQR8 (accessed August 13, 2012).

NAE [National Academy of Engineering]. 2011. Changing the conversation about engineering. *The Bridge* 41(2). Available online at www.nae.edu/File.aspx?id=51068 (accessed August 13, 2012).

NAE Grand Challenge Scholars. 2012. Available online at www.grandchallengescholars.org (accessed June 23, 2012).

Norsk Hydro. 2008a. Caution: Detour ahead. Available at www.hydro.com/en/Press-room/Videos/Train-stunt (accessed August 13, 2012).

Norsk Hydro. 2008b. Taken for a ride. Available online at www.hydro.com/en/Press-room/Videos/Taken-for-a-ride (accessed August 13, 2012).

O'Grady, H. 2010. ACEC New York story: "Your vision can change the world." Available at www.engineeringmessages.org/TakeAction/26035/26039.aspx (accessed June 21, 2012).

Patterson, M. 2010. Lenovo "Engineers Anthem." Available online at http://vimeo.com/10366736 (accessed August 13, 2012).

Sullivan, J. 2010. Changing the conversation at CU–Boulder: A journey to capitalize upon NAE messaging concepts. Available online at www.engineeringmessages.org/23673/26035/26087.aspx (accessed June 20, 2012).

Texas Instruments. 2011. Thank an engineer for MP3 players. Available online at https://sites.google.com/site/thanksengineers/home/videos/thx-classix/thank-an-engineer-for-mp3-players (accessed August 13, 2012).

Vest, C.M. 2011. Engineers: The next generation—Do we need more? Who will they be? What will they do? Address to the 2011 Annual Meeting of the National Academy of Engineering. October 16, 2011. Available online at www.nae.edu/Activities/Events/AnnualMeetings/19611/53074.aspx (accessed August 13, 2012).

Vest, C.M. 2012. Welcome remarks at the NAE Convocation of Professional Engineering Societies. April 16, 2012. Available online at www.nae.edu/File.aspx?id=58735 (accessed August 13, 2012).

Wolsky, M. 2011. Engineering a change in perception: Engineer Your Life and Design Squad. *The Bridge* 41(2): 30–36.

3 CALL TO ACTION

A better public understanding of engineers and engineering will benefit not only the engineering community but the entire country, and to that end the National Academy of Engineering (NAE) has developed messages and taglines for use in engineering outreach. As illustrated in the preceding chapters, the new messaging has been effective in various settings, but its adoption and use are not yet widespread in the engineering community. Broader implementation cannot be achieved by the NAE alone or by the handful of organizations that have taken up the new messaging. It will require the buy-in and participation of a significantly greater portion of the engineering community. And messaging, even if done effectively, is but one contributor to improved public understanding of engineering. Others include opportunity for K–12 students to engage in engineering design activities (e.g., NAE and NRC 2009) and participation of engineers in public affairs, such as through elected office.

In this chapter we lay out an action plan describing basic steps that all segments of the engineering community can take to help change the conversation about engineering. We also describe specific steps for individual segments of the engineering community—companies that employ engineers, professional engineering societies, engineering schools, and others. Because each component has different capabilities and strengths, it makes sense that each should contribute in different ways. The most important thing is that the diverse members of the engineering community begin to speak with one voice, using the general approach outlined in *Changing the Conversation*, when talking about engineering and what it means to the people of this country and the world. We believe outreach by the engineering community will be most effective when it leverages the full range of outreach opportunities and communications media available.

The committee members carefully considered the proposed action items, agreeing that they had to be (1) doable without a major infusion of new financial or human resources and (2) likely to result in measurable improvement in public understanding of engineering. Most of the actions arose from discussions during the project's December 2010 stakeholder workshop, which was specifically designed to elicit suggestions for moving the CTC initiative forward (see agenda in Appendix B); workshop attendees comprised about 40 high-level representatives from across the engineering community (see list of participants in Appendix C).

The committee recognizes that for many organizations there will be tension between maintaining their own brand and implementing broader CTC-based messaging about engineering. This tension is real, but we believe the unique missions of each segment of engineering will more often be enhanced than diminished by steps to improve the public's general understanding of engineering. Industry's bottom line, for example, may be aided by consumers who are more aware of the positive societal impacts of engineering. Engineering schools may find they have a more diverse pool of well-informed applicants to their educational offerings. Engineering professional societies may see an uptick in membership, particularly among younger engineers who want to be associated with an organization that promotes itself as "making a world of difference." And museums and science centers could anticipate larger crowds for their exhibits that highlight the creative problem solving practiced by engineers.

THE BASICS

Organizations and individuals across the engineering community (as well as nonengineers who believe in the importance of messaging) can take several basic actions that will contribute to public understanding of engineering.

The first is to make explicit use of the words "engineer" and "engineering" and express the CTC positioning more frequently in public communications such as press releases, radio and television advertising, websites, social media (e.g., Facebook and Twitter), speeches, and personal email signature blocks. Do not assume that audiences understand "science and technology" to include engineering or that they even understand that the "E" in STEM stands for engineering. This same approach will also work for internal communications—those targeting employees or one's colleagues. As more and more communications efforts become aligned with the CTC messaging, the potential for reaching large swaths of the public improves and the cumulative impact of these seemingly simple actions grows.

The second is to engage more fully with the CTC website (www.engineeringmessages.org) and Facebook page (www.facebook.com/engineersctc). Together, these resources provide practical help for effective messaging and opportunities to build a community of practice. They will have impact only to the degree that organizations and individuals take part. Participation can be as simple as linking to the sites, registering as a member of the CTC website, or liking the Facebook page. It can involve submitting examples of engineering-related messaging, writing case studies of how a particular messaging effort was developed and its impacts, or authoring a blog post. Both sites provide opportunity to express opinions, ask questions, and engage in dialogue.

In addition to these actions, we recommend that the engineering community's efforts to use the CTC messages, taglines, and positioning be based on a view of the engineering profession as a whole rather than specific engineering disciplines. Messaging tuned to individual disciplines may be useful in certain situations, however, and cobranding by combining general and more discipline-specific messaging is certainly possible.

We recommend that all outreach to the public about engineering take into account the special urgency to reach girls, African Americans, Hispanics, and American Indians, groups that data show are significantly underrepresented in engineering in the United States.

Because of their sometimes complementary strengths and abilities to reach different audiences, the various segments of the engineering community should make a more concerted effort to combine forces for more effective messaging. For example, some companies fund research in engineering schools or support programming in science centers and museums. These relationships can be leveraged to expand the reach and impact of engineering messaging. Similarly, engineering professional societies have faculty and student members in many engineering programs around the country, and both engineering schools and societies conduct outreach to K–12 schools. These common interests could be the basis for greater collaboration.

Finally, consistent with the second criterion that guided our choice of action items, we urge organizations that choose to use or adapt the CTC messages to include an impact-assessment component. Ideally, such evaluation should be included up front in the design of the messaging effort rather than as an add-on after the fact. As much as possible, measures of impact should include not only "inputs," such as the number of visits to a website, but also "outputs" that reflect changes in attitudes or behavior—for example, students' views of engineering—as a result of exposure to the CTC messaging.

SECTOR-SPECIFIC ACTIONS

In the effort to get people thinking and talking about engineering in new ways, one of the strengths of the engineering community is the variety of its components—industry, government, professional societies, engineering schools, science and technology centers, and the NAE—each of which typically uses different approaches and targets different groups in their outreach efforts. This diversity can be advantageous as each member of the community takes on different tasks.

Industry

Of all the entities in the engineering community, industry may have the greatest potential to improve public understanding of engineering. It has the resources, the personnel, the expertise, and the motivation to run advertisements or well-organized messaging campaigns prominently featuring engineers. Logically, part of the motivation for doing so would be the desire to improve the engineering pipeline: if more qualified students—including women and underrepresented minorities—can become engineers, the companies that rely on engineering talent will have an easier time finding competent people to hire. Furthermore, many companies could market their products in part by marketing their engineering workforce: "We make the world's best widgets because we have the world's best widget engineers working for us."

Industry's motivation and capacity to improve the understanding of engineering are clear. However, as noted in Chapter 1, efforts by industry have been fragmented and uncoordinated. Different companies promulgate different images of engineers, and the lack of a consistent message dilutes and weakens impact.

Thus the most important thing that industry can do to improve public understanding of engineering is to adopt a single set of messages, such as those described in *Changing the Conversation.* Of course, every company is interested in setting itself apart from others—in promoting its own brand—and that is certainly possible while still conveying a consistent and positive image of engineers and engineering, as illustrated by the examples cited in Chapter 2: Lenovo with its images of engineers in extreme environments, Texas Instruments with its "Thank an engineer" videos, Cisco with its presentation of women engineers discussing how they make people's lives better, and so on.

The committee recommends that industry take the following additional actions to help change the conversation about engineering:

- In the next few years, a main goal for industry should be a significant increase in the number of companies whose corporate identity, recruiting efforts, product advertising, and outreach to the public feature engineers and engineering and use messages and taglines either directly from or comparable to those from *Changing the Conversation.*

- Companies should leverage their outreach and messages by collaborating more often with other segments of the engineering community, such as professional societies and engineering colleges. They could also collaborate among themselves through such mechanisms as the Council on Competitiveness, Business Roundtable, and Change the Equation to create consistent, CTC-based messaging that would reach large segments of the US population.
- Many industries support volunteer outreach by their employees to interest young people in engineering. At DuPont, for example, over 100 employees involved in STEM outreach were trained in the use of the CTC messages; similar training was provided to company staff volunteering at engineering events for the Girl Scouts and for over 20 high schools near the company's headquarters. Industry efforts like these should continue, expand if possible, and be modified to align with the CTC messages.
- Many companies exert a positive influence on diverse segments of society through philanthropic giving. Companies that conduct such activities should ask nonprofit organizations they support and that engage in engineering-related outreach to use the CTC messages.
- Companies that have the resources should consider investing in public service announcements that project a positive image of engineering consistent with the CTC messaging. Such efforts could be tied to television programming that connects to engineering in some way, such as Discovery Channel's Extreme Engineering and MythBusters, the Science Channel's Strip the City, and PBS's Design Squad.

Government Agencies

From the perspective of engineering messaging, there are a number of similarities between industry and certain government agencies, such as NASA, the Department of Defense, National Institute of Standards and Technology, and Department of Energy. All are major employers of engineers, so it is in their interest to improve the diversity and quality of the engineering workforce. The National Science Foundation does not employ large numbers of engineers, but it funds research on engineering and on engineering education, and it promotes the value of science and engineering to the public. The US Patent and Trademark Office, while also not a large employer of engineering talent, plays a key role in evaluating the creative output of engineers and others. All of these agencies can demonstrate how engineering improves lives and shapes the future. Government agencies generally do not spend as much on outreach as large companies, but they do sponsor education outreach activities and communicate with the general public through websites, press releases, visits and presentations by individual employees, and the like.

The committee recommends that federal agencies take the following additional actions to help change the conversation about engineering:

- Education and outreach programs, such as NASA's Summer of Innovation, should incorporate the CTC messages, as should all similar STEM-related government programs that support hands-on experiments and engineering design activities for schools, libraries, scout troops, civic centers, and other organizations.
- Whenever possible, government agencies should collaborate with other segments of the engineering community to advance the goal of changing the conversation—for example, by working with industry partners in outreach programs or regularly participating on the CTC website.
- Training for federally employed engineers who take on speaking and mentoring assignments with students and educators should include exposure to the CTC materials.
- Agencies that provide funding through grants and contracts should create incentives for recipients of this federal support to incorporate CTC messaging in their work.

Professional Societies

Engineering professional societies are different from industry and government agencies because they are member-driven organizations. They can therefore connect with the public on a more personal level, through member visits to classrooms or outreach engagement with other groups. The number of engineering society members is large—several hundred thousand—and they can be mobilized to contribute toward improving public understanding of engineering. With training, these members can be very effective ambassadors for spreading the CTC messages.

The committee recommends that engineering professional societies take the following additional actions to help change the conversation about engineering:

- To provide motivation and vision for working together to improve public understanding of engineering, the societies should develop and endorse an intersociety "memorandum of understanding" to guide their coordinated use of the CTC messages.
- Whether or not they are part of a memorandum of understanding with other engineering professional groups, societies should educate their members about the messages and how to use them. This could be done, for instance, by offering training sessions at society-sponsored conferences and workshops, through webinars (like those sponsored by the National Engineers Week Foundation, described in Chapter 2), and by including articles and editorials about the CTC project in society publications.
- Many societies conduct outreach to teachers, students, and parents. Society-sponsored teacher conferences and workshops could focus on the theme of changing the conversation. And materials provided to K–12 teachers could suggest how engineering makes the world better, reflecting one of the main CTC themes. Communications about the engineering profession as a whole should be the main focus, but messages tuned to individual disciplines are also useful.

American Society for Engineering Education

The American Society for Engineering Education (ASEE) occupies a unique position among engineering professional societies as it does not represent a specific discipline. Rather, its members are primarily engineering educators, and its mission is largely about improving the engineering profession as a whole, not just a narrow slice of it. ASEE's influence extends through the hundreds of engineering and engineering technology education programs around the country, all of which are interested in attracting and retaining students, particularly girls and those from underrepresented groups.

- The committee recommends that ASEE leverage its special connection to engineering educators to broaden their awareness and use of the CTC messages. Specific steps might include creating a recurring session at its annual conference and at the yearly Engineering Deans Council Public Policy Colloquium to review and encourage discussion of efforts to improve engineering messaging in engineering education programs around the country. To bring visibility, ASEE could select one program's messaging efforts each year for special recognition with a one-page spread in *Prism* magazine.

National Engineers Week Foundation

Although not strictly speaking an engineering professional society, National Engineers Week Foundation (NEWF) engages many such societies in coordinated public outreach, particularly through its annual National Engineers Week (E-Week; www.eweek.org). E-Week includes a variety of programs and activities for the general public, some of which incorporate the CTC messages. As noted in Chapter 2, NEWF also developed a webinar-based train-the-trainer module to educate members of the engineering community about the CTC project and how its messages and taglines can be used. The committee recommends that NEWF take the following additional actions to help change the conversation about engineering:

- NEWF should continue to offer web-based, messaging-focused training to engineering student "ambassadors" and volunteers who do outreach to the community and to K–12 schools.
- NEWF's corporate sponsors should be encouraged to incorporate the CTC messaging in their public outreach not only during E-Week but throughout the year.

Engineering Schools

Engineering schools have direct and sustained contact with students—both current and potential—and their parents, as well as teachers and counselors, and can thus influence students' decisions whether or not to pursue engineering as well as their perceptions of engineers and engineering. The efforts at the University of Colorado–Boulder to implement the CTC messages, described in Chapter 2, can be a model for other schools.

The committee recommends that engineering schools take the following actions to help change the conversation about engineering:

- Explain the CTC messaging approach to faculty and staff, describing its rationale and the evidence for its usefulness. This might include new-faculty orientation workshops or other training sessions that lay out explicitly how to use the CTC messages and taglines to shape how students and potential students think about engineering.
- Spread the CTC messages to current and potential future students by, for example, incorporating the CTC messages and taglines in the recruiting and outreach programs of the engineering school. Such efforts need not be directed only to high school juniors and seniors and university students who have not yet decided on a major.
- Work with schools of education so that future K–12 teachers are aware of what engineering is and what engineers do, and encourage schools of education to use CTC-based messages.
- Encourage engineering undergraduates to volunteer in K–12 classrooms, doing engineering design projects and acting as

role models, using CTC-based messages. This kind of outreach can shape K–12 students' impressions of engineers and engineering.

- Aim to educate their own students consistent with the image of engineering outlined in *Changing the Conversation*. If K–12 students are attracted to engineering by the opportunity to both engage their creativity and help people, the engineering curriculum should reflect those qualities. Some schools may wish to use the Grand Challenges for Engineering (www.engineeringchallenges.org) to inspire students to think about problems whose solutions will make a "world of difference." Engineering students educated in these ways can themselves become ambassadors for spreading the CTC messages.

Science and Technology Centers

Science and technology centers have a great deal of contact with the public. But whereas engineering faculty and students, engineers working in industry, and members of engineering professional societies may visit schools and other destinations, science and technology centers are themselves the destination. They are the target of field trips by children from elementary and secondary schools, and they host individuals and families. The Association of Science-Technology Centers (ASTC) reported an estimated 62 million visits to its 345 US member institutions in 2010. Museums such as the Museum of Science in Boston, the Exploratorium in San Francisco, and the Tech Museum in San Jose, California, have developed exhibits and other outreach programs that address engineering issues, but as far as the committee is aware, few if any of these initiatives specifically incorporate the CTC messages.

The committee recommends that science and technology centers take the following actions as part of their contribution to changing the conversation about engineering:

- Insofar as possible, when designing new exhibits or revising existing ones, incorporate the CTC messages. Exhibits and other programming at science and technology centers can

educate the public about engineering in very engaging ways, and this capacity should be leveraged to deliver the CTC messages compellingly and memorably.
- Involve engineers from academia, professional societies, and industry in specific programs and outreach activities that have an engineering or technology focus.

National Academy of Engineering

The NAE occupies a unique position in the engineering community. It has relatively little direct contact with the public (although it is working to change that with the CTC Facebook page), but it fosters close relationships throughout the community. Through its convening powers and published reports, the NAE reaches and sometimes influences the thinking of important decision makers, such as state and federal lawmakers and leaders in industry, government, and academia.

The committee recommends that the NAE take the following additional actions as part of its contribution to changing the conversation about engineering:

- Continue to spread ideas from the CTC project and to promote communication and actions that support public understanding of engineering. This report is the most recent such effort.
- Maintain the CTC website and Facebook page until they are deemed no longer useful. This could be done by the NAE alone or in partnership with other engineering organizations.

OTHER OPPORTUNITIES

There are opportunities for engineering messaging that lie outside the engineering community but nevertheless may be worth pursuing. One is the science- and engineering-focused DIY (do it yourself) movement, in which people build new or modify existing technological devices for amusement or specific practical purposes. The movement is gaining visibility in part through the spread of Maker Faire events, multiple-day festivals celebrating creativity and innovation (www.makerfaire.com). The themes of creativity and innovation are consistent with the CTC messages and taglines, and the festivals' family orientation would allow messaging to reach both children and adults. Funding for the festivals comes largely from corporate donors, many of which are technology-focused companies.

- The committee recommends that organizers of the Maker Faires consider both how engineering messaging can be included in the events themselves and how sponsors can be encouraged to use the messages in their own public outreach. Similarly, the publisher of *MAKE* magazine and its companion website (www.makezine.com) should consider how CTC-related messaging might be infused appropriately in their content.

There is a small but growing movement to introduce engineering into K–12 classrooms in the United States (NAE and NRC 2009). A number of groups have developed curricula for this purpose, and some of these materials have reached national scale. In the coming years, some of the existing curricula will undergo revision, while new curriculum projects will come on line. These efforts may take on new importance with the publication of the *Next Generation of Science Standards* in spring 2013. The standards are expected to include learning goals related to engineering design.

- The committee recommends that developers of K–12 engineering curriculum materials consider using the CTC or similar messages in their materials, promotional efforts, and related teacher professional development.

The biennial USA Science and Engineering Festival, described in Chapter 2, is another potential opportunity for broadening the reach of engineering messages. The NAE and the National Academies have used the messages at these events, but there are hundreds of other participating organizations that could incorporate aspects of the CTC messaging in their festival exhibits as well as in their overall marketing and outreach efforts.

- The committee recommends that leadership of the USA Science and Engineering Festival consider how the messages might be made a more visible component of the event.

CONCLUSION

Like any effort to change public attitudes, improving the engineering "brand" to reflect the creativity of the field and its deep and powerful impacts on society is a long-term proposition. The NAE, through the Changing the Conversation project, has provided new language—a positioning statement, messages, and taglines—and new communication and collaboration tools (the CTC website and Facebook page) in support of this rebranding effort. In the end, of course, it will be up to individuals and organizations in the engineering community itself to take up and use this language and these tools. As our report suggests, the uptake has begun, but much more is possible and needed.

Going forward, members of the engineering community will need to spend more time, more effort, and more resources on consistent, effective engineering messaging. Although the committee does not specifically endorse the idea of a broad national advertisement campaign, comparable to "Got Milk" or similar efforts, such an initiative, with sufficient resources, the right message(s), and strong leadership, would surely be more effective than multiple efforts promulgating inconsistent messages.

As the national and global economy become more technologically sophisticated, it is vitally important to make engineering more understandable, accessible, and appealing to larger and more diverse US audiences—from elementary school students and their teachers and parents, to employers and members of Congress. The action plan described in this report lays out modest but potentially powerful steps for moving toward that goal.

REFERENCES

ASTC (Association of Science-Technology Centers). 2010 Science Center and Museum Statistics. Available online at http://astc.org/about/pdf/Backgrounders/2010%20Science%20Center%20Statistics.pdf (accessed September 19, 2012).

NAE and NRC (National Academy of Engineering and National Research Council). 2009. Engineering in K–12 Education: Understanding the Status and Improving the Prospects. Available online at www.nap.edu/catalog.php?record_id=12635 (accessed January 24, 2013).

A BIOGRAPHIES OF COMMITTEE MEMBERS

Ellen Kullman, *cochair*, is chair of the board and chief executive officer of E. I. du Pont de Nemours and Company, where she has championed market-driven science to drive innovation. She began her career at DuPont in 1988 as a marketing manager and went on to assume positions of increasing responsibility; in 2008 she was tapped to lead the company's focus on growth in emerging international markets. She is a member of the US-India CEO Forum, the Business Council, the executive committee of SCI-America, the board of directors of United Technologies Corp., and the board of overseers at Tufts University School of Engineering. She is also a member of the board of Change the Equation (CTEq), a national coalition of more than 100 CEOs committed to improving science, technology, engineering, and mathematics (STEM) learning for US pre-K–12 students. She holds a BS in mechanical engineering from Tufts University and a master's in management from Northwestern University.

Charles M. Vest (NAE), *cochair*, is president of the National Academy of Engineering (NAE) and president emeritus of the Massachusetts Institute of Technology. He began his career at the University of Michigan, where he taught in the areas of heat transfer, thermodynamics, and fluid mechanics and conducted research in heat transfer and engineering applications of laser optics and holography before turning his attention to academic administration. He served as associate dean and then dean of engineering and then provost and vice president for academic affairs. In 1990 he became president of MIT and served in that position until December 2004. As president of MIT, he was active in science, technology, and innovation policy; building partnerships among academia, government, and industry; and championing the importance of open, global scientific communication, travel, and sharing of intellectual resources. During his tenure, MIT launched its OpenCourseWare (OCW) initiative; cofounded the Alliance for Global Sustainability; enhanced the racial, gender, and cultural diversity of its students and faculty; established major new institutes in neuroscience and genomic medicine; and redeveloped much of its campus. Vest has served on various federal committees and commissions, including the President's Committee of Advisors on Science and Technology (PCAST) during the Clinton and Bush administrations, the Secretary of Education's Commission on the Future of Higher Education, and the Secretary of State's Advisory Committee on Transformational Diplomacy. He earned a BS in mechanical engineering from West Virginia University in 1963 and MSE and PhD degrees in mechanical engineering from the University of Michigan in 1964 and 1967, respectively.

G. Wayne Clough (NAE) is secretary of the Smithsonian Institution, leading the world's largest museum and research complex with 19 museums, 9 research centers, the National Zoo, and research activities in more than 90 countries. Before his appointment to the Smithsonian, he was president of the Georgia Institute of Technology for 14 years. Clough has been a professor at Duke University, Stanford University, and Virginia Tech. He also served as head of the Department of Civil Engineering and dean of the College of Engineering at Virginia Tech and as provost at the University of Washington. He has received nine national awards from the American Society of Civil Engineers (ASCE), including the 2004 OPAL lifetime award for contributions to education, and has twice been awarded civil engineering's oldest recognition, the Norman Medal, in 1982 and in 1996. He received the George Westinghouse Award from the American Society for Engineering Education in 1986 for outstanding teaching and research and the National Engineering Award from the American Association of Engineering Societies in 2002, and in 2008 he was recognized with the NAE Bueche Award for his efforts in public policy. He is a member of the National Science Board and served on the President's Council of Advisors on Science and Technology (2001–2008). He was cochair of the 2004 National Innovation Initiative, university vice chair of the US Council on Competitiveness, and chair of the NAE's Engineer of 2020 project. Clough's interests include science, technology, and higher-education policy, sustainability, international programs, museums, and history. He has published more than 130 papers and reports and six book chapters and has cowritten numerous committee reports. He received his bachelor's and master's degrees in civil engineering from Georgia Tech in 1964 and 1965 and a doctorate in 1969 in civil engineering from the University of California, Berkeley.

Leslie Collins is executive director of the National Engineers Week Foundation, where she is responsible for development and stewardship of more than 150 corporate, government, engineering society, education, and diversity partners representing more than 1.5 million engineers and engineering students. She initiated the DiscoverE K–12 volunteer program for Engineers Week 1990; National Engineers Week Future City Competition in 1993; Introduce a Girl to Engineering Day in 2001; Global Marathon For, By and About Women in Engineering in 2005; and the foundation's Diversity Council in 2008. She began her career in public relations at the American Gas Association and went on to become public relations director for the National Society of Professional Engineers (NSPE). She is a graduate of Boston College and attended graduate school at the University of Fribourg in Switzerland.

Don P. Giddens (NAE) is dean emeritus of the College of Engineering at the Georgia Institute of Technology, where he is a professor in the Wallace H. Coulter Department of Biomedical Engineering at Georgia Tech and Emory University. He is also Georgia Research

Alliance Eminent Scholar Emeritus and president of the Parabola Group, LLC, in Atlanta. He joined the Georgia Tech faculty in 1968, after 2 years in the aerospace industry. In 1992 he left his position as chair of Aerospace Engineering to serve as dean of the Whiting School of Engineering and professor of mechanical engineering at the Johns Hopkins University until 1997, when he rejoined Georgia Tech to establish the Wallace H. Coulter Department of Biomedical Engineering. He is past president and fellow of the American Society for Engineering Education (ASEE) and a fellow of the Biomedical Engineering Society (BMES), American Institute for Medical and Biological Engineering (AIMBE), ASME, and American Association for the Advancement of Science (AAAS). He chaired the NAE committee that authored the report *Changing the Conversation: Messages for Improving Public Understanding of Engineering*. Giddens has served in a variety of professional activities involving engineering education and biomedical research and remains active and consults on various issues associated with engineering education, diversity in engineering, and biomedical engineering research. He received all his degrees (BAE 1963, MSAE 1965, and PhD 1966) from Georgia Tech.

Robert Hoffman is president and chief creative officer of Hoffman and Partners, an advertising and marketing firm based in Boston, Massachusetts. His work combines business discipline with creative problem solving. After several years of domestic and international marketing efforts in the music industry, he teamed with business associate Dan Gearon to found Gearon Hoffman in 1982. The firm achieved national visibility for successful branding efforts on behalf of Rockport footwear, the *New York Daily News*, Prudential Retirement Services, Massachusetts General Hospital, Grolsch beer, and Boston's Museum of Science. Gearon Hoffman has also assisted a number of academic institutions, including Worcester Polytechnic Institute and Simmons College. A student of human behavior, Hoffman is sensitive to what drives consumers to act; as a writer, strategist, and creative director, he explores the emotional links that connect consumers with brands. He has a BA in English from Antioch College and pursued graduate studies in communication at the University of Pennsylvania.

Ray O. Johnson, a global executive focused on business, innovation, and diversity, is senior vice president and chief technology officer of the Lockheed Martin Corporation, where he provides leadership in the strategic areas of technology, engineering, production operations, and program management. He also leads the corporation's Advanced Technology Laboratory and the Center for Innovation, a world-class laboratory for collaborative experimentation and analysis that involves customers and industry partners. He has experience in strategic planning, program development, program management, and venture capital funding. He is a member of the governing board of the Indo-US Science and Technology Forum and a sponsor of the DST–Lockheed Martin India Innovation Growth Program. He is also a member of the board of visitors for the A. James Clark School of Engineering and the School of Computing, Mathematics, and Natural Sciences at the University of Maryland, the Dean's Advisory Council for the College of Engineering at Carnegie Mellon University, and the board of affiliates of the Rice University Professional Science Master's Program. Johnson also chairs the USA Science and Engineering Festival's Advisory Board and the US Council on Competitiveness Technology Leadership and Strategy Initiative. He holds a BS in electrical engineering from Oklahoma State University and an MS and PhD in electrical engineering from the Air Force Institute of Technology.

Virginia (Gini) Kramer is executive creative director at the advertising and public relations firm Keiler & Company, where she oversees print and broadcast advertising, collateral, direct mail, and interactive projects. She is an award-winning copywriter who has broad experience working with clients in a variety of industries: financial services, banking, insurance, health care, aerospace, high technology, medical devices, pharmaceuticals, manufacturing, publishing, and the performing arts. She participated in the NAE's 2005 preliminary focus group on public understanding of engineering messaging and served on the committee that wrote *Changing the Conversation: Messages for Improving Public Understanding of Engineering.* She graduated (magna cum laude) from the University of Hartford.

Irving Pressley McPhail is president and chief executive officer of the National Action Council for Minorities in Engineering (NACME). He joined NACME in 2007 as executive vice president and chief operating officer (COO), after having served 15 years as a college president or chancellor at the Community College of Baltimore County, St. Louis Community College at Florissant Valley, and LeMoyne-Owen College. He was also COO of the Baltimore City Public Schools. Under his leadership, the Community College of Baltimore County was named one of 12 Vanguard Learning Colleges in the United States and Canada in 2000 by the League for Innovation in the Community College, and in 2003 was awarded the PBS O'Banion Prize for Leading the Way to Change in Teaching and Learning. Working at the nexus of practice, policy, and research in literacy education, postsecondary student success, community college leadership, and STEM education, McPhail is the coeditor of *Teaching African American Learners to Read: Perspectives and Practices* (2005) and the author of more than 50 journal articles, chapters, monographs, and technical reports. He serves on the board of directors of the Society of Manufacturing Engineers Education Foundation and is a charter member of the E-Week [Engineers Week] Diversity Council. He also served as a strategy coach in the Achieving the Dream national initiative at Prairie View A&M University in Texas. He earned a bachelor's degree in development sociology at Cornell University and a master's in reading at the Harvard Graduate School of Education and was awarded an honorary doctor of engineering degree by the Polytechnic Institute of New York University.

E. James (Jim) Prendergast is executive director and COO of IEEE, leading a staff of approximately 1,100 employees in US and overseas locations. He was previously corporate vice president and chief technology officer for DuPont Electronic and Communication Technologies, where he accelerated growth in the electronics and communications high-tech markets, and before that he was vice president and director of Motorola's Physical Sciences Research Laboratories, where he directed long-range research in future integrated systems, energy, lab-on-a-chip, and speculative "reach-out" initiatives. He received a PhD in electrical engineering from Cambridge University.

Betty Shanahan is executive director and CEO for the Society of Women Engineers (SWE), a 24,000-member organization that establishes engineering as a highly desirable career aspiration for women. She is also a champion of diversity in engineering and the engineering profession overall. Before joining SWE in 2002, she spent 24 years in development, engineering management, and marketing for the electronics and software industries. She has been responsible for products and technologies in document viewing and conversion, parallel processing, signal processing, and computer-aided software engineering. Most recently she was vice president of product management and marketing for the Software Components Division of Stellent, Inc. She is president of the Council of Engineering and Scientific Society Executives and in 2012 was named one of STEMConnector's "100 Women Leaders in STEM." She is also a certified association executive and a member of the IEEE, ASCE, and ASME. She earned a BS in electrical engineering from Michigan State University, a master's in software engineering from the Wang Institute of Graduate Studies, and an MBA in strategic management from the University of Chicago Booth School of Business.

B WORKSHOP AGENDA

National Academy of Engineering
Keck Center of the National Academies
500 5th St., NW
Washington, D.C.

November 30–December 1, 2010

TUESDAY, NOVEMBER 30

9:00 a.m. **Welcome and Plans for the Workshop**
Ellen Kullman, Dupont; Chuck Vest, NAE

9:15 a.m. **Remarks from the Sponsor**
Tom Peterson, NSF

9:30 a.m. **Introduction of Participants**
Around the table

10:00 a.m. **Overview of *Changing the Conversation* Report and Current Project**
Don Giddens, Georgia Tech; Greg Pearson, NAE

10:30 a.m. **Break**

10:45 a.m. **Examples of the Use of CTC Messaging**
Panel Moderator: Leslie Collins, Committee Member

- Jackie Sullivan, College of Engineering and Applied Sciences, University of Colorado–Boulder (student recruitment materials)
- Julie Benyo, WGBH Television (Engineer Your Life website)
- Nital Patel, Insight Technology, Inc. and IEEE ("Engineers Make a World of Difference" student video contest)

12:30 p.m. **Working Lunch: Demonstration and Discussion of CTC Website**
Derek Rector, Diamax Information Systems Corporation

1:30 p.m. **Rebranding Engineering: Challenges and Opportunities**
Mitch Baranowski, BBMG
Respondent: Gini Kramer,Committee Member

3:00 p.m. **Reflection/Discussion: Mixed-Sector Breakout Groups**

4:30 p.m. **Breakout Group Reports and Discussion**
Group Rapporteurs
Discussion Moderator: Ellen Kullman, Committee Cochair

5:30 p.m. **Day 1 Adjournment**

WEDNESDAY, DECEMBER 1

8:30 a.m. **Reflections on Day 1**
All participants
Moderator: Chuck Vest, NAE

Proposed focus question:

1. What steps will be needed, by whom, to move these ideas forward?

9:30 a.m. **Benchmarking: How Other Industries and Professions Address the Communications Challenge**
Maria Ivancin, Market Research Bureau

10:00 a.m. **Messaging Actions: Idea Generation**
All participants

10:30 a.m. **Sector-Specific Breakout Groups**
Facilitator: Betty Shanahan, Committee Member

11:30 a.m. **Breakout Group Reports**

11:45 a.m. **Final Thoughts**
Chuck Vest, NAE

12:00 p.m. **Workshop Adjournment**

C PARTICIPANTS LIST

Mitch Baranowski
Principal & Chief Creative Officer
BBMG

Julie Benyo
Director, Educational Outreach
WGBH Television

Bob Black
Deputy Executive Director
American Society for Engineering Education (ASEE)

Jim Buczkowski
Director, Global Electrical and Electronics Systems Engineering
Ford Motor Company

Doug Coffey
Vice President of Executive Communications
BAE Systems, Inc.

Leslie Collins
Executive Director
National Engineers Week Foundation

Rick Dastin
President, Enterprise Business Group
XEROX

Wayne Davis
Dean, College of Engineering
University of Tennessee

Mark Dean
IBM Fellow & Vice President
IBM

Lizabeth Fogel
Director of Education
The Walt Disney Corporation

James Foster
Director of Process Development
Archer Daniels Midland Company

Gayle Gibson
Director, Corporate Operations
DuPont

Don Giddens
Dean, College of Engineering
Georgia Institute of Technology

Avi Gopstein
Senior Advisor
Office of the Under Secretary for Science
US Department of Energy

Kathryn Gray
President 2005–2006
National Society of Professional Engineers

Elizabeth Grossman
Technology Strategy and Policy
Microsoft Corporation

Maria Ivancin
President
Market Research Bureau LLC

Louisa Koch
Director, Office of Education
National Oceanic and Atmospheric Administration (NOAA)

Virginia Kramer
Executive Creative Director
Keiler & Company

Ellen Kullman, *Cochair*
Chair of the Board and CEO
DuPont

Mark Little
Senior Vice President & Director, Global Research
General Electric

Tom Loughlin
Executive Director
American Society of Mechanical Engineers

Pender McCarter
Senior Public Relations Counselor
IEEE-USA Washington

Chris McManes
Public Relations Manager
IEEE-USA Washington

Deirdre Meldrum
Dean, Ira A. Fulton Schools of Engineering
Arizona State University

Richard Miller
President
Olin College of Engineering

Pat Natale
Executive Director
American Society of Civil Engineers

David Oberbillig
Einstein Fellow
US Department of Energy

Rafaela Ornelas-Schwan
Director of Development
Great Minds in STEM

Nita Patel
Vice President for Communications/Public Awareness
IEEE-USA Washington

Greg Pearson
Senior Program Officer
National Academy of Engineering

Paul Peercy
Dean, College of Engineering
University of Wisconsin, Madison

Mary Petryszyn
Vice President, Mergers and Acquisitions
Raytheon Integrated Defense Systems

Darryll Pines
Dean, College of Engineering
University of Maryland

Robert Pool
Consultant
Digital Pens LLC

Oscar Porter
Executive Director
Math Engineering Science Achievement

Derek Rector
President
Diamax Information Systems Corporation

Phil Reed
Department of STEM Education & Professional Studies
Old Dominion University

Elizabeth Shanahan
CEO & Executive Director
Society of Women Engineers

Michael Smith
Deputy Executive Director, Internal Operations
National GEM Consortium

Stephen Smith
Director, Technical Activities & Communications
American Institute of Chemical Engineers

Kendall Starkweather
Executive Director, International Technology and Engineering Educators Association

James Stofan
Deputy Associate Administrator for Program Integration
NASA Office of Education

Jacquelyn Sullivan
Associate Dean, College of Engineering and Applied Science
University of Colorado

John Tracy
Chief Technology Officer & Senior Vice President of Engineering
Boeing Company

Charles M. Vest, *Cochair*
President
National Academy of Engineering